GOVERNANCE OF TENURE **TECHNICAL GUIDE** No. 1

Governing land *for* women and men

A technical guide to support the achievement of responsible gender-equitable governance land tenure

FOOD AND AGRICULTURE ORGANIZATION OF THE UNITED NATIONS
ROME, 2013

Reprinted January 2013

This publication is intended to support the use of the Voluntary Guidelines on the Responsible Governance of Tenure of Land, Fisheries and Forests in the Context of National Food Security. It is not intended to contradict the language of the Guidelines as endorsed by the Committee on World Food Security on 11 May 2012 nor the role of states in their implementation.

ISBN 978-92-5-107403-9

Acknowledgements

This technical guide on *Governing land for women and men* has been prepared by a team led by Elizabeth Daley and including Clara Mi Young Park, Francesca Romano, Leila Shamsaifar, Ting-Hui Lau and Babette Wehrmann. The guide was drafted by Elizabeth Daley and Clara MiYoung Park, with other contributors of material and content including Judy Adoko, Dalal Alnaggar, Marianna Bicchieri, Patricia Costas, Kate Dalrymple, Nigel Edmead, Debra Anne Fletcher, Seema Gaikwad, Catherine Gatundu, Renée Giovarelli, Paolo Groppo, Nora Heil, Adriana Herrera-Garibay, Rachael Knight, Annalisa Mauro, Eliane Najros, Esther Obaikol, Martha Osorio, Sabine Pallas, Neil Pullar, Peter Rabley, Rebecca Sittie, Elizabeth Stair, Josephine Stowers-Fiu, Mika-Petteri Törhönen, Margret Vidar and Babette Wehrmann.

The guide benefited from comments from Dubravka Bojic, Paolo Groppo, Paul Munro-Faure, Ana Paula de la O Campos, Martha Osorio, David Palmer, Neil Pullar, Francesca Romano, Margret Vidar and Babette Wehrmann. It has been externally peer reviewed by Bayarmaa Byambaa, Catherine Gatundu, Renée Giovarelli, Susana Lastarria-Cornhiel, Sabine Pallas, Siraj Sait, Agatha Wanyonyi and the German Working Group on Land hosted by the German Agency for International Cooperation (GIZ). The editing of the guide was by Jane Shaw and the layout by Luca Feliziani.

FAO thanks the International Fund for Agricultural Development (IFAD) for funding the development and preparation of this technical guide.

Foreword

On 11 May 2012, the Committee on World Food Security endorsed the *Voluntary Guidelines on the Responsible Governance of Tenure of Land, Fisheries and Forests in the Context of National Food Security* (the Guidelines). Based on the principles of sustainable development and in recognition of land's centrality to development, these Guidelines are intended to contribute to global and national efforts towards the eradication of hunger and poverty by promoting secure tenure rights and equitable access to land, fisheries and forests.

This technical guide on *Governing land for women and men* aims to assist implementation of the Guidelines' principle of gender equality through the achievement of responsible gender-equitable governance of land tenure.

Preparation of this guide included substantive research and consultation, leading to the production of an FAO working paper in February 2011, *Governing land for women and men – gender and voluntary guidelines on responsible governance of tenure and other natural resources* (FAO, 2011a). The working paper analysed and identified the main themes and issues to be covered by this guide.

To help identify good practice for responsible gender-equitable governance of land tenure, a technical workshop was held in May 2011 at FAO Headquarters in Rome. "Governing Land for Women and Men – a Workshop on Gender-Equitable Governance of Land Tenure" brought together land sector practitioners from government, civil society and the private sector around the world to share practical, hands-on experiences and approaches to improving gender equity in land tenure governance. The outcomes of this workshop have contributed immensely to enriching this guide.

After endorsement of the Guidelines, the drafting team finalized this technical guide through a process that included peer review.

Alexander Müller
Assistant Director-General
FAO Natural Resources Management and Environment Department

Contents

MODULE 4:
Technical issues

MODULE 5:
Getting the message across

Resources and references

Introduction

LIST *of* ILLUSTRATIONS

Boxes

Tables and figures

Introduction
Gender-equitable governance of land tenure

"States should consider the particular obstacles faced by women and girls with regard to tenure and associated tenure rights, and take measures to ensure that legal and policy frameworks provide adequate protection for women and that laws that recognize women's tenure rights are implemented and enforced. … States should develop relevant policies, laws and procedures through participatory processes involving all affected parties, ensuring that both men and women are included from the outset."

(*Voluntary Guidelines on the Responsible Governance of Tenure of Land, Fisheries and Forests in the Context of National Food Security*, **from paragraphs 5.4 and 5.5**).

What is this guide about?

This technical guide on *Governing land for women and men* aims to assist implementation of the *Voluntary Guidelines on the Responsible Governance of Tenure of Land, Fisheries and Forests in the Context of National Food Security* (FAO, 2012b) by providing guidance that supports the Guidelines' principle of gender equality in tenure governance. At the beginning of each module, reference is made to the relevant provisions in the Guidelines.

The guide focuses on equity and on how land tenure can be governed in ways that address the different needs and priorities of women and men. It moves away from long-standing debates about gender equality in access to land, towards the mainstreaming of gender issues to achieve more gender-equitable participation in the processes and institutions that underlie all decision-making about land.

Gender equality is one of the ten core principles for implementation of the Guidelines and is closely tied to three others – human dignity, non-discrimination, and equity and justice (Box 1).

Gender relations determine differences in the relative status and power of women and men. Gender and gender roles shape the opportunities and constraints that women and men face in securing their livelihoods, including their access to land, decision-making about land, and

BOX 1:
The Guidelines' principles of implementation relevant to gender equity and equality

Source: FAO, 2012b.

3.B.1 Human dignity: recognizing the inherent dignity and the equal and inalienable human rights of all individuals.

3.B.2 Non-discrimination: no one should be subject to discrimination under law and policies as well as in practice.

3.B.3 Equity and justice: recognizing that equality between individuals may require acknowledging differences between individuals, and taking positive action, including empowerment, in order to promote equitable tenure rights and access to land, fisheries and forests, for all women and men, youth and vulnerable and traditionally marginalized people, within the national context.

3.B.4 Gender equality: Ensure the equal right of women and men to the enjoyment of all human rights, while acknowledging differences between women and men and taking specific measures aimed at accelerating de facto equality where necessary. States should ensure that women and girls have equal tenure rights and access to land, fisheries and forests, independent of their civil and marital status.

the institutions that administer land and other productive resources (FAO, 2003a).

Gender-equitable governance of land tenure ensures that women and men can participate equally in their relationships to land, through both formal institutions and informal arrangements for land administration and management (Table 1).

TABLE 1:
Definitions

TERM	DEFINITION
Gender	The social and political identity attached to a person's biological sex and physical body – what it means to be a woman or a man, which may differ in different places. Gender relations are the relations between women and men, based on the social identity attributed to each sex within a society (MacKinnon, 1987; Moore, 1994).
Land tenure	The relationships between people and land, defined legally or customarily – the rules invented by societies to allocate property rights over land, grant access to the use and control of land, and define the associated responsibilities and restraints. Land tenure systems determine who can use what resources for how long and under what conditions (FAO, 2002b; 2007).
Governance	The process of governing – managing society and reconciling competing priorities and interests of different groups. It includes formal government institutions and informal arrangements, and is concerned with how citizens participate in decision-making, how government is accountable to its citizens, and how society obliges its members to observe its rules and laws (FAO, 2007; 2009b).

What are the scope and limitations of this guide?

Gender analysis helps to identify the needs and priorities of women and men, clarifying the relevance of gender across different age, wealth, caste, race, ethnic, religious and other groups (FAO, 2003a). However, this level of detailed analysis is not systematically possible in a guide such as this one, which instead aims to be broadly applicable and relevant to all the national and local government officials, civil society groups, administrators, technicians and professionals working in the land sector worldwide.

This guide focuses on gender-equitable governance of land tenure; it therefore does not embrace the full scope of the Guidelines, which also cover governance of tenure of fisheries and forests. However, the broad principles of gender-equitable participation to governance processes are applicable to fisheries and forests as well (particularly as these interface with land) and interesting lessons can be drawn for wider use.

The guide provides advice on mechanisms, strategies and actions that can be adopted to improve gender equity in the processes, institutions and activities of land tenure governance. It does not include an in-depth discussion of all the outcomes of these processes and institutions, but provides useful recommendations as basic guidance.

Why a technical guide on gender-equitable governance of land tenure?

Gender equity has to be mainstreamed into the governance of land tenure to ensure that the interests and needs of all people – women and men – are addressed, and that the processes and institutions for land tenure are transparent, consultative and participatory.

Mainstreaming of gender equity in land tenure governance is also essential for achievement of Millennium Development Goals (MDGs) 3 – promoting gender equality and empowering women; and 1 – eradicating poverty and hunger. As The State of Food and Agriculture 2010–2011 argues, gender inequality must be addressed to increase the overall global performance of the agricultural sector (FAO, 2011b). Mainstreaming is supported by firm internationally endorsed principles of non-discrimination by gender, as most clearly set out in the 1979 Convention on the Elimination of all Forms of Discrimination against Women (CEDAW) (Box 2).

Other important international and regional instruments that uphold principles of non-discrimination by gender and are relevant to the pursuit of gender-equitable governance of land tenure can be found in the *Resources and references* section at the end of this guide.

BOX 2:
CEDAW provisions supporting gender-equitable governance of land tenure

Articles 13 to 15: Women to have equal rights with men to conclude contracts and in the administration of property, including equal access to mortgages and equality in procedures before courts and tribunals, with restrictions to the legal capacity of women deemed to be null and void.

Article 14: Women to receive equal treatment in land and agrarian reforms, and in land resettlement schemes and housing.

Article 16: Women to have the same rights in marriage as their spouses with respect to ownership, management, enjoyment and disposal of property, including on divorce.

What key issues do the modules cover?

This guide consists of five modules, plus this introduction and a concluding section listing resources and references. Table 2 outlines the themes and contents of these modules.

Gender-equitable participation in land policy-making processes is a first step towards responsible gender-equitable governance of land tenure. All implementation activities should therefore be rooted in the core principles of human dignity, non-discrimination, equity, justice and gender equality. *Module 1* addresses gender-equitable participation in land policy-making processes, considering the issues for different sets of stakeholders and different levels of governance.

A second step is translating policy into law. In many countries, the presence of different tenure systems creates confusion when different laws and customary norms and practices conflict. In other cases, good land laws may not be applied because regulations, procedural manuals and implementation strategies are lacking. *Module 2* addresses legal issues, such as the drafting of laws, access to justice, land dispute resolution and legal support.

TABLE 2:
Contents of this guide

MODULE	THEME	SYNOPSIS
1	**Policy-making**	Building gender-equitable participation into land policy-making processes
2	**Legal issues**	Legal pluralism, legal drafting, access to justice, land dispute resolution, legal support
3	**Institutions**	Gender-equitable participation and representation in the institutions of land tenure governance
4	**Technical issues**	Gender equity in land administration activities, technologies and systems, and women's participation in land administration
5	**Getting the message across**	Communication strategies and methods to support responsible gender-equitable land tenure governance

Women are often excluded from participation in the day-to-day processes of land tenure governance at all levels, and have limited capacity to influence decision-making. *Module 3* addresses strategies for facilitating the representation and participation of all women and men in the institutions of land tenure governance – including customary institutions – and for developing the capacities of these institutions.

Technical issues of land administration also need gender-sensitive approaches, methods and technologies. *Module 4* addresses two issues: ***i)*** how adequately land administration activities are designed to take into account and address women's needs, interests and concerns; and ***ii)*** how women can participate equally with men in land administration activities, as target groups and clients.

Good communications are vital to responsible gender-equitable governance of land tenure. *Module 5* considers the most effective communication strategies and methods and addresses key issues in getting the message across, including gender sensitization, awareness raising, advocacy, legal literacy and long-term change in values and attitudes.

The concluding section provides sources of further guidance and information. Users of this guide wishing to improve their understanding of the analytical issues are also encouraged to refer to the wider literature on gender and land. FAO's *Governing land for women and men – gender and voluntary guidelines on responsible governance of tenure and other natural resources* provides a background analysis of all the themes and issues covered in this guide, including obstacles and bottlenecks to achieving gender-equitable land tenure governance (FAO, 2011a).

Who is this guide for, and how should it be used?

This guide is a reference tool that provides administrators, technicians and professionals working in the land sector with guidance and examples of good practice – what has worked, where, why and how – for achieving land tenure governance that is gender-equitable.

It also aims to inspire a broader audience, including anyone working in any capacity on issues relating to gender and/or land.

Although the modules in the guide follow a logical sequence – from policy, to law, to implementation – different users in different situations may take different modules as their starting point for introducing change, and each module may be more or less relevant to different types of stakeholder.

Each module is thus designed as a stand-alone section that can be consulted and used separately or in combination with others. Figure 1 indicates which of the five modules may be most relevant for different types of stakeholder.

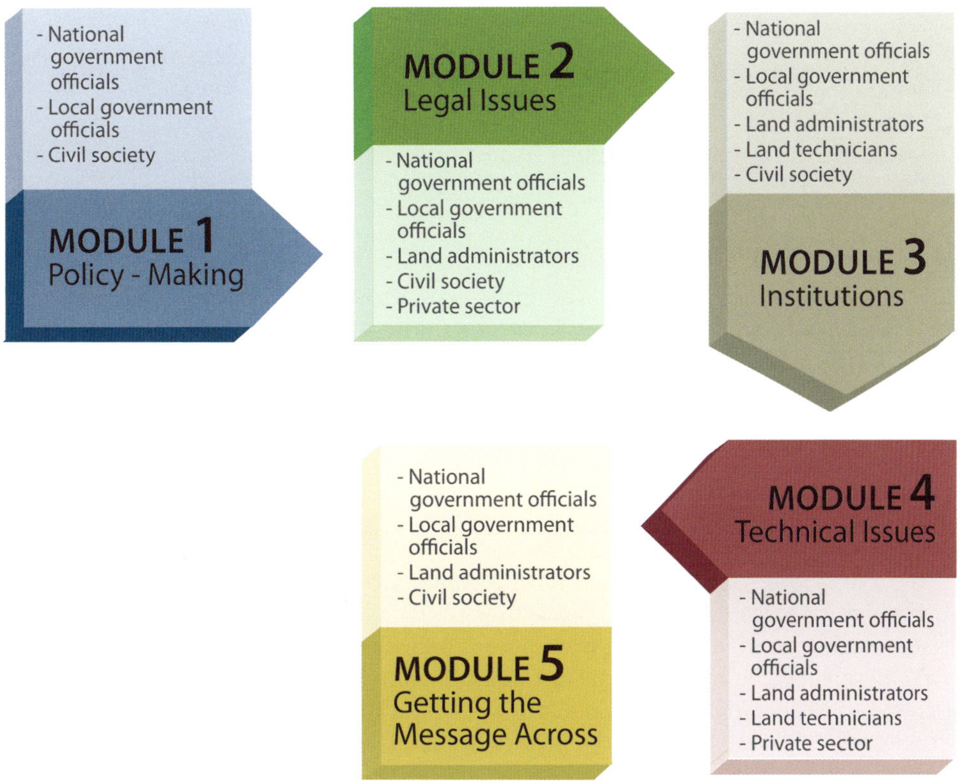

FIGURE 1:
Modules and stakeholders

The material presented is drawn from existing literature and from approaches and experiences in the field. The information sources used in preparing each module are listed in the references at the end of the module; a comprehensive list of references is also given in the section on *Resources and references*.

FAO. 2002b. *Land tenure and rural development. Land Tenure Studies 3.* Rome. ftp://ftp.fao.org/docrep/fao/005/y4307e/y4307e00.pdf

FAO. 2003a. *SEAGA Socio-Economic and Gender Analysis Programme macro level handbook.* Rome. www.fao.org/docrep/012/ak229e/ak229e00.pdf

FAO. 2007. *Good governance in land tenure and administration. Land Tenure Studies No. 9.* Rome. ftp://ftp.fao.org/docrep/fao/010/a1179e/a1179e00.pdf

FAO. 2009b. *Towards voluntary guidelines on responsible governance of tenure of land and other natural resources – discussion paper.* Land Tenure Working Paper No. 10. Rome. ftp://ftp.fao.org/docrep/fao/012/i0955e/i0955e00.pdf

FAO. 2011a. *Governing land for women and men – gender and voluntary guidelines on responsible governance of tenure of land and other natural resources*, E. Daley and C.M. Park. *Land Tenure Working Paper No. 19.* Rome. http://www.fao.org/docrep/014/ma811e/ma811e00.pdf

FAO. 2011b. *State of Food and Agriculture 2010–11: Women in agriculture, closing the gender gap for development.* Rome. www.fao.org/docrep/013/i2050e/i2050e.pdf

FAO. 2012b. *Voluntary Guidelines on the Responsible Governance of Tenure of Land, Fisheries and Forests in the Context of National Food Security.* Rome. http://www.fao.org/docrep/016/i2801e/i2801e.pdf

MacKinnon, C. 1987. *Feminism unmodified: Discourses on life and law.* Cambridge, Massachusetts, USA, Harvard University Press.

Moore, H. 1994. The divisions within: sex, gender and sexual difference. *In A passion for difference – essays in anthropology and gender.* Oxford, UK, Polity Press.

MODULE

1

Policy-making

LIST *of* ILLUSTRATIONS

Boxes

Checklists

Tables and figures

MODULE 1: **Policy-making**

Gender-equitable participation in land policy-making

Introduction[1]

The key to ensuring responsible gender-equitable governance of land tenure is to start with a gender-equitable policy-making process in which all stakeholders, women and men, are equally included in formulating and implementing land policies.

However, power differences between women and men, as well as other factors such as caste, race, ethnicity or age, often create hierarchies that determine who participates in policy-making, affecting the likelihood of gender-equitable processes and outcomes.

Gender-equitable participation in land policy-making follows directly from one of the ten core principles of implementation of the Guidelines – consultation and participation (Box 1.1).

BOX 1.1:
Policy-making and the Guidelines

Source: FAO, 2012b.

A) Provisions on land policy-making with reference to gender issues

3.B.6 Consultation and participation: engaging with and seeking the support of those who, having legitimate tenure rights, could be affected by decisions, prior to decisions being taken, and responding to their contributions; taking into consideration existing power imbalances between different parties and ensuring active, free, effective, meaningful and informed participation of individuals and groups in associated decision-making processes.

5.5 States should develop relevant policies, laws and procedures through participatory processes involving all affected parties, ensuring that both men and women are included from the outset. Policies, laws and procedures should take into account the capacity to implement. They should incorporate gender-sensitive approaches, be clearly expressed in applicable languages, and widely publicized.

9.10 State and non-state actors should strive, where necessary, together with representative institutions of affected communities and in cooperation with affected communities, to provide technical and legal assistance to affected communities to participate in the development of tenure policies, laws and projects in non-discriminatory and gender-sensitive ways.

B) Provisions on the contents of land policies with reference to gender issues

5.3 States should ensure that policy, legal and organizational frameworks for tenure governance recognize and respect, in accordance with national laws, legitimate tenure rights including legitimate customary tenure rights that are not currently protected by law; and facilitate, promote and protect the exercise of tenure rights. ... States should provide frameworks that are non-discriminatory and promote social equity and gender equality.

10.1 Where informal tenure to land, fisheries and forests exists, States should acknowledge it in a manner that respects existing formal rights under national law and in ways that recognize the reality of the situation and promote social, economic and environmental well-being. States should promote policies and laws to provide recognition to such informal tenure. The process of establishing these policies and laws should be participatory, gender sensitive and strive to make provision for technical and legal support to affected communities and individuals.

Distinguishing between land policies and governance in land policy-making

Before considering the policy-making process itself, it is important to note that the process is different from the content of the resulting land policies, which may not all be gender equitable to the same degree.

However, in any country, a gender-equitable land policy-making process with the participation of women and men of all classes, castes, ethnicities, races and religions allows different perspectives on all land issues to be discussed and considered. An inclusive policy-making process is therefore more likely to result in more gender-equitable land policies.

Land policies can encompass a wide range of land-related issues, all with potential implications for responsible governance of land tenure. Gender-equitable land policies vary from country to country, according to how land rights are dealt with. Policy actions for achieving gender-equitable governance of land tenure include:

- applying international conventions on women's rights;
- recognizing women's customary rights to land;
- making sure that women can register their land rights under the same conditions as men;
- making sure that all stakeholders can participate equitably in land tenure governance, including in local land administration and management institutions;
- introducing positive discrimination measures.

Building gender-equitable participation into land policy-making processes

Building gender-equitable participation is a long-term process that requires planning and interventions at various levels, and the involvement of different groups of stakeholders. Political will on the part of government is needed, along with engagement and support by stakeholders in civil society, the private sector and international organizations.

Figure 1.1 indicates five practical elements in building a process based on the equal participation of women and men. These elements are interrelated and will usually need to be pursued at the same time.

FIGURE 1.1: **Building a gender-equitable land policy-making process**

Context analysis

Each country has its own unique set of geographical, economic, social and cultural features, as well as its own political system and patterns of land use and landholding. All stakeholders need to understand this local context so they can participate in the policy-making process as effectively as possible. For example, in many countries legal pluralism is part of the context. This means that policy-makers should consider communal and customary tenure arrangements, the links between land tenure and the rules governing property rights within marriage and on inheritance, and the hierarchies of power that affect decision-making regarding land.

Sometimes there are similarities in context among the countries within a region, such as in Asia and Africa (Box 1.2). Such similarities allow the sharing of lessons learned, directly between countries or through regional initiatives such as the African Union's Framework and Guidelines on Land Policy in Africa, which is described in Box 1.5.

BOX 1.2:
Regional similarities in the context for land policy-making

Sources: Seema Gaikwad's presentation on "Asian experiences of civil society engagement with land policy processes" and Catherine Gatundu's presentation on "Africa experiences of civil society involvement in land policy process" at the May 2011 workshop; IPC, 2010; FAO, 2008b.

Asia

A coalition of civil society organizations established Land Watch Asia to support, monitor and advocate for access to land, agrarian reform and equitable and sustainable rural development across Asia. Land Watch Asia builds on regional similarities in:

- the declining availability of land for agriculture;
- a post-colonial history of conflicts and civil strife;
- the impact on women's rights to land of the multiple burdens of class, caste and gender discrimination, sexual abuse and violence:
  - family laws provide differently for women's rights to individual and marital property and inheritance, depending on ethnicity, religion or caste;
  - women generally have few and unrecognized rights to land and are excluded from decision-making and governance processes;
  - gender inequality and biases have historically constrained women's access to government initiatives to register land and rationalize property rights.

Africa

The LandNet West Africa network comprises civil society organizations and activists engaged in land policy-making processes in their respective countries. It builds on regional similarities in:

- agriculture's importance in national economies;
- women's important role in food production;
- the pressures of desertification, drought and population growth;
- the colonial legacy of many land laws but few guiding policies;
- a predominance of customary land tenure arrangements;
- the insecurity of land tenure faced by many women:
  - local governance institutions and structures are often dominated by men, and customary rules on land are often interpreted in ways that marginalize women;
  - women are disadvantaged by their greater poverty, lower literacy levels and lack of knowledge about the legal system;
  - land administration practices often disadvantage women, including in registering household land and acquiring strong individual rights to it.
- differences in the rights of citizenship enjoyed by men and women – especially married women – which affect women's relationship to all types of property; women sometimes need the consent of their fathers or husbands to register land that they have acquired themselves.

CHECKLIST 1.1:
Context analysis

✓	How are different groups of women and men usually included in political debate and discussion within local communities and at the national level?
✓	Do cultural and social norms require that participatory processes be separate for women and men, and for other groups? For example, do women or youth feel constrained about speaking openly in meetings where adult men are present? Is it locally acceptable for women and men to mix openly in public meetings?
✓	Do women and men belong to separate social institutions and fora, which can be harnessed for consultations on land policies?
✓	Do cultural and social norms mean that women and men have different types and subjects of knowledge about land?

REMEMBER!
Proper context analysis at the outset of a process will ensure adoption of the most effective approach and combination of strategies for that context. What works in one country may not work equally well in another, even though lessons can be learned and shared.

Advocacy

Advocacy based on evidence demonstrates the value of a gender-equitable process, which benefits from the contributions and participation of all stakeholders, both women and men, making it much more likely to result in consensus and support for the resulting policies. Potential audiences for advocacy – policy-makers, researchers and people at the grassroots – need to be identified, contacted and integrated into the development and dissemination of communication products, strategies and methods, as described in *Module 5*.

CHECKLIST 1.2:
Advocacy

Source: Sabine Pallas's presentation on "Developing communication and advocacy tools through action-oriented research in Africa" at the May 2011 technical workshop.

✓	Gather as much evidence as possible to back up the arguments.
✓	Establish who is likely to be affected by the research, and include them in the process.
✓	Involve local actors, both women and men, taking into account differences in status and power that may affect the likelihood that they will participate.
✓	Present the evidence in ways that policy-makers will understand, without distorting the findings to fit the existing policy agendas of different stakeholders.
✓	Identify who is advocating on behalf of whom and who is mandated by whom, to avoid situations in which an organization advocates for people without adequately consulting them about their needs and demands.

REMEMBER!
Actors undertaking advocacy work at the national level may have different agendas from those at the grassroots level. Alignment and harmonization of advocacy goals is essential to ensure that the advocacy work serves the interests of all.

Civil society organizations often have an important role in advocacy (Checklist 1.5).

Sensitization and training

Training and sensitization on gender and land will be much more productive than training on women's land issues alone, and will help to avoid marginalization and misconception of gender issues as being only about women. Women and men politicians and government officials will require training to become effective land policy-makers who represent the issues of women and men equally.

Ordinary women and men may need special training and support to be able to participate in land policy-making processes. Particular attention should be paid to facilitating women's participation when gender relations in the country present challenges. For example, women-only meetings enable women to discuss sensitive land issues among themselves before addressing these issues within the broader land policy-making process. Civil society organizations – particularly women's groups – have a special role in facilitating the creation of inclusive women's networks, and local government and customary institutions at the grassroots can include women-only fora.

Multi-stakeholder dialogue

Policy-making is one of the core functions of government, so land policies and the land policy-making process must ultimately be owned by government. To facilitate constructive engagement and alliance-building, other stakeholders have to accept this ownership and to understand the competing demands on land and the overall government positions on land policy issues and on gender. However, cooperation and participation of all stakeholders should start from the outset of the policy-making process, and continue throughout the development, adoption and implementation phases, including in the drafting of gender guidelines and programme design.

Long-term approach

Most policy-making processes are long enough to allow lessons to be learned along the way. Box 1.3 provides an example from Uganda.

BOX 1.3:
Lesson learning for gender-equitable land policy-making

Uganda's Constitution of 1995 provides equality between women and men, including in the acquisition and holding of land, and affirmative action to remedy the historical discrimination faced by women and other marginalized groups. However, confusion during preparation of the 1998 Land Act resulted in the omission of an important clause on co-ownership of land between women and men. In 2004, the act was amended to require the consent of both spouses for transactions involving family land, but civil society activists still felt that the law's provisions for women's access, control and ownership of land were inadequate.

Lessons have been learned, and Uganda's new Land Policy, before the Ugandan Cabinet in July 2011, included women's groups in the policy-making process under the umbrella of the Uganda Land Alliance. These groups participated in a review of 16 land-related laws to make sure that gender is mainstreamed in all aspects of landownership, access, use, administration and management in Uganda.

Sources: WLLA, 2010a; Esther Obaikal, personal communication.

Practical steps for a gender-equitable land policy-making process[2]

Each of the following four steps contains aspects relating to the five key elements illustrated in Figure 1.1 – context analysis, advocacy, sensitization and training, multi-stakeholder dialogue, and long-term approach.

In step 1, an internal government mandate in support of a gender-equitable, participatory policy-making process is developed and pursued by relevant government departments and ministries. Civil society and international organizations can provide information, awareness raising and capacity development to key government officials, especially potential "internal champions" (see step 4), and can help to translate the internal mandate into actions. It is very important that government is open to consulting both women and men, but the country context is also important. Where gender equality has further to progress, it may be possible to ensure only the participation of women who are involved in the government. Building an internal mandate and implementing actions related to it take time and must be adequately resourced.

In step 2, a four-way multi-stakeholder alliance of government, civil society, the private sector, and international organizations is developed to promote gender-equitable participation in land policy-making. This involves a mixture of top-down and bottom-up engagements, activities and events throughout the policy process, incorporating both women and men. Civil society organizations and women's groups can form direct alliances with private sector stakeholders (such as land administrators and land technicians), and work with international organizations to lobby governments. The final goal is a broad and consensual multi-stakeholder alliance involving all stakeholders.

In step 3, the multi-stakeholder alliance seeks to establish and train a group of "pioneers" to spearhead policy debate and discussion at the local level, taking into account the different contexts across the country. This pioneer group will have a multiplier effect as the land policy-making process becomes more widely known. It should adapt local activities and events to make them gender-equitable, by taking into account such issues as language, the timing of meetings and the most appropriate forms of engagement. The pioneer group should include subgroups of women and men in each part of the country. A core group should visit the provinces, regions, districts and counties to build local alliances involving both women and men and empowering them to develop their own capacities and alliances at the grassroots level. These alliances will be valuable for local-level land policy-making.

In step 4, "champions" from government, civil society and the private sector are identified and engaged to ensure that gender-equitable participation continues throughout the land policy-making process. Affirmative action and quotas for policy-making institutions and structures may be considered, along with ways of making policy-making more consultative and participatory for all the public. The media have an important role in providing information and raising awareness.

CHECKLIST 1.3:
Improving gender equity in land policy-making

✓	Efforts to include women in land policy-making processes and to address their specific concerns and issues should be clearly targeted and adequately resourced.
✓	Women and other marginalized groups are often the most resource-poor and time-constrained participants in policy-making. Where cultural constraints prevent women from talking openly, they should be consulted separately from men.
✓	The structure and format of the policy-making process must be flexible enough to accommodate women's time constraints and family/cultural obligations.
✓	Even where women's participation is legally or formally incorporated into land policy-making processes, attention should be paid to ensuring their actual engagement.
✓	Quotas or similar mechanisms to ensure women's participation in policy-making are an important starting point, but women may also need training and support for active and constructive engagement, and help to become policy-makers in the first place.

REMEMBER!
It can be difficult to measure progress towards gender-equitable participation in land policy-making. It may be tempting to examine only the resulting land policy instead. However, it is important to make sure that women and men are also equally involved in the policy-making process.

Stakeholders' roles in gender-equitable land policy-making

Governments

In providing gender-equitable land policy-making processes governments should become active in two areas:

- Government officials must be sensitized on gender issues, including gender-equitable participation in land policy-making, as part of the induction training for all new staff in relevant ministries and departments. Where politicians take part directly in government administration, they too should be sensitized.
- Governments can encourage other stakeholders' participation in the land policy-making process, and set the overall agenda and mandate for an inclusive, participatory and gender-equitable process. Box 1.4 provides examples of government leadership in Ghana, Nepal and India.

BOX 1.4:
Government support for gender-equitable land policy-making

In *Ghana*, the government's Land Administration Project (LAP) included the development of a gender strategy for addressing the land administration concerns of women and men through gender-sensitive data gathering, and participation in the design, implementation, and monitoring and evaluation processes. The strategy has five thrusts for achieving LAP's gender objectives: public education, capacity building, institutional processes, advocacy, and collaboration and networking.

In *Nepal*, a representative of the National Land Rights Forum – which comprises women and men farmers, women's organizations and organizations working on land reforms – sits as a member of the government's seven-member Nepal Land Reform Commission.

In *India*, in 2008, the government set up a Committee of State Agrarian Relations on the Unfinished Task of Land Reform to examine the status of various land reforms, including measures to ensure women's participation in land governance processes and greater access to land. Specific recommendations included women's representation in the agencies set up to monitor land reforms.

Sources: Seema Gaikwad's presentation on "Asian experiences of civil society engagement with land policy processes" and Rebecca Sittie's presentation on "Gender issues in managing and implementing land registration in Ghana" at the May 2011 technical workshop; India Ministry of Rural Development, 2009; Ghana Land Administration Project, no date.

CHECKLIST 1.4:
Activities for government

✓	Use affirmative action such as quotas to ensure that women are equitably represented in the political and governance institutions and structures involved in land policy-making. Consider capacity development and flexibility to support these institutions and structures.
✓	Establish a group of stakeholders or a specialist government agency to address gender issues by: ‣ raising awareness about the land policy-making process and land policy issues among women and men; ‣ gathering international good practice and lessons for gender-equitable participation in land policy-making from neighbouring countries.
✓	Establish and train a pioneer group of government officials, including women, to spearhead the land policy-making process at the local level, by visiting districts and other decentralized government institutions to: ‣ identify contextual variations for women and men countrywide; ‣ establish and train groups of local women and men to take this process to the grassroots.
✓	Take the lead in adapting policy activities, events, processes and tools to ensure gender-equity, for example by conducting an internal audit of gender-equitable participation.

REMEMBER!
Governments play the lead role in gender-equitable policy-making, but they need to work with other stakeholders to ensure that policies reflect people's needs and to help generate consensus and support for the policy outcomes.

Civil society

Civil society stakeholders can help make marginalized people and groups more visible through: advocating for their issues and supporting their direct participation in the policy debate; monitoring and reporting on the policy-making process to make it more accountable on gender equity, such as by using radio to inform the public on how women and men are being engaged; monitoring and ensuring the accountability of government and private sector stakeholders; and developing and leading broad public consensus for gender-equitable participation in land policy-making.

Many countries lack research and knowledge on land issues. As they are close to the grassroots, civil society stakeholders can undertake systematic field-level research to help ensure that women's voices, knowledge and interests are included in land policy-making; private sector stakeholders and international organizations can fund this research. For example, between 2008 and 2010, the International Land Coalition (ILC) coordinated a research project on securing women's access to land in Eastern and Southern Africa, with the specific aim of influencing the formulation and implementation of land policies through advocacy based on the research results. The project involved multi-disciplinary research teams and fostered partnerships among researchers, non-governmental organizations (NGOs) and grassroots organizations (see WOLREC, 2011 and ILC's Web site on Securing Women's Access to Land for further examples).[3]

The extent to which civil society stakeholders, including women's organizations, can engage in land policy-making processes depends on:

- the political space available for civil society in the country;
- the relationship between civil society and the government, and the government's attitude to civil society involvement in land policy-making;

- the level of community knowledge and awareness about the land policy-making process;
- the length of the policy-making process (including technical delays), which may put pressure on civil society resources.

A challenge for civil society engagement in land policy-making is ensuring the participation and representation of women and men from all social groups, as explained in more detail in *Module 3*. This challenge is exacerbated when issues other than gender equity in land policy-making are the priority.

Box 1.5 provides examples of civil society's important roles in formulating and implementing gender-equitable land policies. Civil society stakeholders are well placed for carrying out advocacy, mobilizing communities and rallying the public's support for and participation in change, while engaging directly with governments; and they can provide training, legal and other support to other stakeholders. As one women's organization notes, it is important for civil society "to have a clearly defined role, in order to work in partnership with the Government while at the same time maintaining independence" (WLLA, 2010b: 9).

BOX 1.5:
Civil society support for gender-equitable land policy-making

Africa

In *Africa*, civil society has engaged persistently and constructively in formal policy-making processes:

- During preparation of the *African Union*'s Framework and Guidelines on Land Policy in Africa, civil society stakeholders participated in regional consultation meetings, commented on draft papers, and developed parts of the text. Civil society groups also organized side-events at African Union meetings during the policy-making process, and issued land policy communications.
- In *Kenya*, during formulation of the land policy, civil society groups under the Kenya Land Alliance mobilized communities to demand space in the policy-making arena. More than 1000 women and men were directly engaged in an open process involving public fora and six thematic groups, including one addressing gender equity. The new land policy was adopted in December 2009. Civil society groups also contributed to constitutional debates, including by proposing the constitutional principles for land reform that underlie the new land policy. The new Constitution, adopted in 2010, gives women and men the right to equal treatment and opportunities in political, economic, cultural and social spheres; ensures that elective bodies are made up of at least one-third women (and/or at least one-third men); and eliminates gender discrimination in inheritance and access rights to land and property.

Asia

In *Asia*, civil society stakeholders have invested in formal and customary governance structures that promote inclusivity, transparency and accountability, and have engaged in land policy reform:

- In *India*, in 2004, women's organizations launched a movement to remove gender discriminatory provisions in the Hindu Succession Act of 1956. The amended act of 2005 gives daughters the same rights to partnership as sons. In 2008, the Committee of State Agrarian Relations on the Unfinished Task of Land Reform recommended mandatory joint entitlement and ownership rights to homestead lands, and community rights for women, including ownership over common property land in villages.
- In *Nepal*, in 2009 and 2011, landless women held sit-ins, rallies and meetings with high-ranking government officials. In response, the government formed a seven-member task force, including a representative of the National Land Rights Forum, to look into recommendations made by the High-Level Scientific Land Reform Commission and to ensure women's right to land.
- Also in *Nepal*, women's organizations and legal aid groups have provided elected women representatives with training and capacity development so they can participate effectively in land policy-making processes and voice their opinions on issues around women's land rights.

Sources: Catherine Gatundu's presentation on "Africa experiences of civil society involvement in land policy process", Judy Adoko's presentation on "Constructive ways of working with customary institutions in support of women's land rights", and Seema Gaikwad's presentation on "Asian experiences of civil society engagement with land policy processes" at the May 2011 workshop; ULA, 2010; Land Watch Asia, 2010.

CHECKLIST 1.5:
Actions for civil society

Source: Catherine Gatundu's presentation on "Africa experiences of civil society involvement in land policy process" at the May 2011 technical workshop.

Formulation	
✓	Engage directly with government.
✓	Propose text options and prepare background and issues papers.
✓	Use the media to change attitudes, build support and create public consensus for the participation of both women and men in an inclusive and consultative policy-making process, including by disseminating policy content and technical documents in ways that everyone can understand (a major strength of civil society).
✓	Provide legal back-up and support.
✓	Organize side-events on policy-making processes at meetings, conferences and other events.
Implementation	
✓	Organize and participate in debates on the development of legislative frameworks.
✓	Train staff and members of land administration institutions on the gender aspects of new land policies.
✓	Campaign for implementation of land policies, for example by setting up watchdog groups, and by using score-cards to track government's implementation progress, and publicizing the findings.
✓	Monitor the gendered impacts of land and property policies and legislation.

REMEMBER!
Civil society is most effective when it works in cooperation with all other stakeholders, including the private sector and international organizations that may provide funding for civil society's policy-making work.

BOX 1.6:
Government and civil society working together

Source: Patricia Costas's presentation on "Gender-equitable communication methods and sensitization strategies to support land governance in Bolivia" at the May 2011 technical workshop.

Bolivia's Agrarian Reform Law of 1953 contained a non-discrimination clause but permitted only single women, including widows, to become beneficiaries of land distribution and titling. The law establishing the National Service for Agrarian Reform in 1996 brought equity for all women and men, regardless of marital status, and led to increased land titling for women. In 2006, a new Constitution improved the legal framework for women's rights to land, and the government committed itself to ensuring women's full participation in land reform decision-making.

Women's organizations have been central protagonists throughout these processes. In 1990, NGOs such as the National Federation of Peasant Women were involved in marches for territory and dignity for indigenous women. In 2006, they participated in the constituent assembly, for the first time in Bolivian history. An alliance of women's organizations and NGOs set out a legislative agenda for women, including the incorporation of women's property rights into the new Constitution. These successes involved an actively engaged civil society providing the research and evidence to support change, and have helped to create a positive environment for gender-equitable land tenure governance in the country.

The private sector

The private companies and practitioners carrying out much of the day-to-day technical work in the land sector often have specialist knowledge and insights that can be of value in developing a gender-equitable land policy. They also often have access to government officials, and already influence the outcomes of land policy-making processes through the technologies and systems they use and promote. Private sector initiatives have become an increasing part of land and agricultural reform across the globe. They include initiatives for redistributive reforms, as in the sugar industry in

South Africa, and innovative mechanisms for affordable land registration processes through partnerships with micro-finance institutions, as in Ghana (Bernstein, 2005; Rabley, 2010).

However, while government stakeholders are accountable through the political system, and civil society stakeholders are accountable to their funders, the role of the private sector within the wider land policy-making process must be made more transparent. At present, private sector initiatives are often not quantified or even officially tracked. Accountability of private sector stakeholders is especially important because they are usually powerful actors with strong corporate interests. Accountability therefore rests on promoting and monitoring corporate social responsibility.

Private sector stakeholders can sign up to international guidelines promoting social responsibility and gender equity, against which their engagement in the policy-making process can be measured and assessed in any country. The Guidelines provide a vital set of principles that all private sector stakeholders engaging in land policy-making should embrace and commit to.

International organizations

International organizations are often major sponsors and donors, providing funds to support land policy-making efforts in many countries. Their responsibility to support gender equity stems directly from their commitments to the Guidelines and to the range of international instruments that uphold principles of gender equity and equality. The United Nations (UN) System-Wide Action Plan (SWAP) on Gender Equality and the Empowerment of Women[4] was adopted in 2012 to measure gender equality across the UN system and to further the achievement of gender equality and women's empowerment within UN policies and programmes.

International organizations also have roles in mediating between the government and civil society and private sector stakeholders, and in ensuring the active cooperation of all these stakeholders in the policy-making process. This role should be used to promote and advocate for gender equity.

Where international organizations, including international NGOs and bilateral donors, provide direct support and funding to civil society organizations engaging in gender-equitable land policy-making, the media and other stakeholders have a responsibility to ensure accountability through informing citizens about the involvement of these international stakeholders in their country's land policy-making process.

Local-level land policy-making

Intra-country variations by province, district and even community can influence the effectiveness of stakeholders' participation in land policy-making processes, activities and events. Decentralization allows these local variations to be taken into account, and facilitates the development of capacity at lower levels of governance for gender-equitable policy-making on many of the day-to-day aspects of land administration and management. The pioneer group described in step 3 of the section on practical steps is vital in helping to establish and develop capacity at decentralized levels, particularly by concentrating on the involvement and participation of women.

Issues to bear in mind when planning capacity development at this level are the lower education and literacy levels that often exist in rural areas, and the need to use local languages and simple explanations. Gender-equitable land policy-making at the local level is closely linked to the communication and awareness-raising issues addressed in *Module 5*.

Summary of key messages on policy-making

Equitable participation of all stakeholders, women and men, is necessary.

The land policy-making process is different from, but strongly influences, the content of the resulting policies.

An understanding of the local context is vital for effective participation.

Sensitization and training on gender equity – and on land issues for both women and men – are needed at all levels.

Governments own the land policy-making process, but all stakeholders should cooperate and participate from the outset. Multi-stakeholder dialogue helps build support for the resulting policy outcomes.

Land policy-making is a long and challenging process and lessons can be learned along the way.

Important steps in a gender-equitable land policy-making process include building an internal mandate, developing multi-stakeholder alliances, establishing a pioneer group, and finding national and local champions.

Ongoing gender sensitization of the government officials involved in land policy-making is essential.

Civil society's roles in advocating for gender-equitable participation, monitoring land policy-making.

The private sector should be more openly and transparently included in land policy-making, and held accountable on gender equity issues as part of corporate social responsibility.

International organizations have a responsibility to promote gender equity when supporting and engaging in land policy-making efforts.

The construction of a gender-equitable land policy-making process should be a participatory, deliberate and concerted effort involving all stakeholders at all levels, and it must be adequately resourced.

Suggested monitoring indicators for policy-making

- % of women and men members of formal land policy-making institutions and structures
- % of formal land policy-making meetings and fora that include participation from grassroots groups representing both women and men
- % of government officials directly involved in the land policy-making process who have been trained and sensitized on gender issues
- % of women and men among government officials participating in land policy-making structures and institutions
- % of women and men among civil society representatives participating in land policy-making structures and institutions
- % of women and men among private sector representatives participating in land policy-making structures and institutions
- % of international organization funding for the land policy-making process allocated to the promotion of gender-equitable participation in policy-making

Notes

[1] Dalal Alnaggar, Kate Dalrymple, Patricia Costas, Catherine Gatundu, Seema Gaikwad, Paolo Groppo, Rachael Knight, Annalisa Mauro, Rebecca Sittie, Elizabeth Stair and Sabine Pallas provided substantial inputs to this module during the May 2011 technical workshop at FAO Headquarters, including through discussions of the policy-making working group.

[2] This section draws on discussions of the policy-making working group at the May 2011 workshop.

[3] www.landcoalition.org/global-initiatives/womens-land-rights/swal

[4] www.unwomen.org/wp-content/uploads/2012/05/swap.pdf

Bernstein, A. 2005. *Land reform in South Africa – a 21st century perspective. Research Report No. 14.* Johannesburg, Centre for Development and Enterprise.

FAO. 2008b. *Good governance and natural resources tenure in South East Asia region,* O. Nabangch ang and E. Srisawalak. *Land Tenure Working Paper No. 4.* Rome.

FAO. 2012b. *Voluntary Guidelines on the Responsible Governance of Tenure of Land, Fisheries and Forests in the Context of National Food Security.* Rome. http://www.fao.org/docrep/016/i2801e/i2801e.pdf

Ghana Land Administration Project. no date. *An overview of LAP's gender initiative.*www.ghanalap.gov.gh/index1.php?linkid=154&sublinkid=161

India Ministry of Rural Development. 2009. *Committee on State Agrarian Relations and Unfinished Task of Land Reforms Vol. 1 – Draft Report.* New Delhi, Government of India.

IPC. 2010. *Civil Society Consultation on FAO's Voluntary Guidelines on Responsible Land and Natural Resources Tenure, Kuala Lumpur, Malaysia, 24–26 March 2010, Final Report.* Rome, International Planning Committee for Food Sovereignty.

Land Watch Asia. 2010. *Nepal: renewed push for land reform as a new chapter in history unfolds. Campaign Update 2010–2011.* www.angoc.org/portal/wp-content/uploads/2012/06/05/land-watch-asia-campaign-updates/nepal_final.pdf

McAuslan, P. 2010. Personal reflections on drafting laws to improve women's access to land: Is there a magic wand? Journal of Eastern African *Studies,* 4 (1): 114–130.

Rabley, P. 2010. Alleviating poverty in the developing world – leveraging property rights with geospatial technology. *The American Surveyor, January 2010.* www.amerisurv.com/pdf/theamericansurveyor_rabley-alleviatingpoverty_January2010.pdf

ULA. 2010. *The Uganda Land Alliance quarterly newsletter, Vol. 4.* Kampala.

WLLA. 2010a. *The impact of national land policy and land reform in Uganda.* Accra, Women's Land Link Africa.

WLLA. 2010b. *The impact of national land policy and land reform on women in Zambia.* Accra.

WOLREC. 2011. *Women's access to land and household bargaining power: a comparative action research project in patrilineal and matrilineal societies in Malawi. Research Report No. 9.* Rome, ILC, Women's Legal Resource Centre. www.landcoalition.org/sites/default/files/publication/959/wlr_9_malawi.pdf

MODULE

2

Legal issues

LIST *of* ILLUSTRATIONS

Boxes

Checklists

Tables and figures

MODULE 2: **Legal issues**

The foundations for achieving gender-equitable governance of land tenure

Introduction[5]

The law provides the foundations for achieving responsible gender-equitable governance of land tenure. Gender-equitable laws and rules provide an enabling environment (Englert and Daley, 2008) in which women and men are more likely to be able to exercise and realize their rights to land equitably.

The importance of legal issues is linked to two of the ten core principles of implementation of the Guidelines – rule of law and accountability (Box 2.1).

BOX 2.1: **Legal issues and the Guidelines**

Law making

3.B.7 Rule of Law: adopting a rules-based approach through laws that are widely publicized in applicable languages, applicable to all, equally enforced and independently adjudicated, and that are consistent with their existing obligations under national and international law, and with due regard to voluntary commitments under applicable regional and international instruments.

3.B.9 Accountability: holding individuals, public agencies and non-state actors responsible for their actions and decisions according to the principles of the rule of law.

Gender-related content of laws and procedures

4.4 Based on an examination of tenure rights in line with national law, States should provide legal recognition for legitimate tenure rights not currently protected by law. Policies and laws that ensure tenure rights should be non-discriminatory and gender sensitive. ... All forms of tenure should provide all persons with a degree of tenure security which guarantees legal protection against forced evictions that are inconsistent with States' existing obligations under national and international law, and against harassment and other threats.

4.6 States should remove and prohibit all forms of discrimination related to tenure rights, including those resulting from change of marital status, lack of legal capacity, and lack of access to economic resources. In particular, States should ensure equal tenure rights for women and men, including the right to inherit and bequeath these rights.

5.3 States should ensure that policy, legal and organizational frameworks for tenure governance recognize and respect, in accordance with national laws, legitimate tenure rights including legitimate customary tenure rights that are not currently protected by law; and facilitate, promote and protect the exercise of tenure rights. ... States should provide frameworks that are non-discriminatory and promote social equity and gender equality.

↓

Source: FAO, 2012b.

9.2 Indigenous peoples and other communities with customary tenure systems that exercise self-governance of land, fisheries and forests should promote and provide equitable, secure and sustainable rights to those resources, with special attention to the provision of equitable access for women. Effective participation of all members, men, women and youth, in decisions regarding their tenure systems should be promoted through their local or traditional institutions, including in the case of collective tenure systems. Where necessary, communities should be assisted to increase the capacity of their members to participate fully in decision-making and governance of their tenure systems.
10.1 Where informal tenure to land, fisheries and forests exists, States should acknowledge it in a manner that respects existing formal rights under national law and in ways that recognize the reality of the situation and promote social, economic and environmental well-being. States should promote policies and laws to provide recognition to such informal tenure. The process of establishing these policies and laws should be participatory, gender sensitive and strive to make provision for technical and legal support to affected communities and individuals.
Dispute resolution from a gender perspective
21.1 Dispute resolution services should be accessible to all women and men, in terms of location, language and procedures.
25.3 Where appropriate, States may consider using customary and other local mechanisms that provide fair, reliable, gender-sensitive, accessible and non-discriminatory ways of promptly resolving disputes over tenure rights to land, fisheries and forests.

Legal pluralism from a gender perspective

Legal pluralism is common throughout the world and refers to a situation where several different types of legal regimes apply to the same territory; for example, where there are customary, statutory and religious land laws, or different tenure arrangements for publicly and privately held land. This makes for a complex legal framework of overlapping rights, competing authorities and often contradictory rules, with potentially negative implications for gender-equitable governance of land tenure when laws and customary norms and practices conflict (FAO, 2011a: 12).

Understanding of the context and complexities of the legal framework in the country concerned is vital. There is need to consider not only land laws, but also family and inheritance laws, marriage laws, civil and rural codes, as well as customary and religious laws and rules, and the ways in which they interact and/or overlap. For example, family law can have a significant impact on individual rights to land – in some countries property rights within marriage may be determined according to whether the marriage ceremony was civil, customary or religious.

Box 2.2 illustrates some of the ways in which legal pluralism is relevant to responsible gender-equitable governance of land tenure in different countries.

BOX 2.2
Multiple tenure arrangements and legal pluralism

In *India*, the Hindu Succession (Amendment Act) of 2005 provides for equal inheritance regardless of gender, but the Muslim Personal Law does not, so people of different religions are treated differently.
In *Bolivia*, the 2006 Constitution provides for a Ministry of indigenous, traditional and rural justice to connect statutory and customary justice systems. However, although indigenous and women's land rights are protected, there is no specific legal protection for indigenous women's land rights, and little intersection between the protection of the two sets of rights.

In *Uganda*, the split between statutory and customary law is addressed in the Constitution, which provides for equality between women and men in all spheres, declares decisions that deny women rights to land null and void, and upholds equality between spouses at all stages of marriage. However, although the Land Act requires spousal consent for transactions affecting joint matrimonial property, the marriage has to be formally registered for this provision to apply, so people married under customary law are not protected.

In *Mozambique*, statutory land law is gender-equitable and has clear provisions on equal rights for women and men. Decisions about land are generally based on customary law, as long as it is consistent with the Constitution and the Land Law, including provisions on gender equity. Under customary law, however, women obtain land rights through their husbands or fathers, and the State has not established a mechanism for monitoring compliance with the Constitution and the Land Law to ensure that women's statutory rights are upheld. In addition, customary law is often misused, and most rural women do not know that statutory land law protects their rights.

In *Kyrgyzstan* and *Tajikistan*, although national laws recognize women's rights to land, housing and property, customary law and traditional patriarchal approaches often prevail. Since the collapse of the Soviet Union, Islamic principles are re-emerging. Islamic principles stipulate women's rights to land and property and, when enforced, provide some degree of tenure security for women. However, these principles are not always entirely applied: for instance, men may opt to marry in the Muslim way - without registration - but do not follow Islamic principles within the marriage, and especially during divorce. Growing numbers of rural marriages are not formally registered, so do not fall under the protection of the formal law, while Muslim law offers no legal protection because there are no tools for enforcing it. Women, therefore, cannot claim any property in cases of divorce or when their husbands die or abandon them.

Sources: Nakayi, 2010; WLLA, 2010a; Daley, 2011: 41–42; Undeland, 2012; Marianna Bicchieri, Patricia Costas, Seema Gaikwad and Rachael Knight personal communication.

The links between land and property laws and the family laws that relate to marriage and inheritance are particularly relevant to tenure reforms. In many countries, customary marriage and inheritance rules lead to the perpetuation of gender inequalities in land relations across generations (Guyer, 1987). Any tenure reform that does not take this into consideration will have limited impact.

Responsible gender-equitable governance of land tenure thus requires that land and family laws are reformed together (FAO, 2011a: 17; McAuslan, 2010: 125). A recent successful example comes from Rwanda, where land tenure reforms followed reforms to the rules on inheritance and matrimonial property and have had some successful gender outcomes, with women participating in local land committees and registering their land (Daley, Dore-Weeks and Umuhoza, 2010). The Rwanda Government's application of quotas throughout the political system also helped to ensure women's participation.

Customary law and land tenure practices evolve in line with a changing legal environment, changing social and economic conditions, and external factors such as climatic, demographic and financial shocks. Box 2.3 illustrates this evolution as experienced by Western Australia's indigenous Aboriginal people.

BOX 2.3:
Gender and indigenous land rights

Australia's Aboriginals, who comprise 3.1 percent of Western Australia's population of 2.3 million people, have always had a very strong spiritual attachment to the land and strongly gendered knowledge about land. Traditionally, responsibility for land was patrilineal, and knowledge relating to land still resides largely in men, especially senior men. Women's knowledge is limited to women-only sites largely related to food collection, birthing and so on.

The 1993 Native Title Act established a clear legal process for Aboriginal groups to claim and register native land titles, and legally recognizes their right to participate in decision-making about land.

Source: Debra Fletcher's presentation on "Land administration – improving gender-equity in institutions and activities: Aboriginal participation in land management in Australia with emphasis upon Western Australia" at the May 2011 technical workshop.

However, Aboriginal land claims are highly contested and legal recognition of native land titles depends on proof based on the patrilineal traditions and customs. In the meantime, however, Aboriginal law and custom have changed to a cognatic system, which recognizes descent from the mother as well as the father. Aboriginal women, and young men, have assumed greater responsibilities in their communities, including as spokespeople and decision-makers in some land settlements, which have a broader governance base than the traditional decision-making structures centred around senior men. These changes to law and custom have strengthened women's claim to the right to be involved in land management. Through case law, judges are broadening what is accepted as proof for land claims – adapting the law to reflect changes in gendered information and knowledge, such as by legally recognizing cognatic descent in certain cases.

Although women and men in Western Australia's Aboriginal communities will retain their different roles in land management for the foreseeable future, they are transferring knowledge about land and land claims in flexible ways in response to changing gender roles.

Power dynamics at the community level are another factor to be taken into account in ensuring gender-equitable governance of land tenure within legal pluralism. Local power dynamics enable some people to use/abuse the multiple land tenure arrangements and institutions of legal pluralism and customary law at the expense of other people, particularly vulnerable groups such as widows. For instance, in-laws or agnates (relatives on the paternal side) may be able to dispossess widows or female relatives (such as sisters) by claiming ownership using formal land titles that are in men's names, or by using local customary or statutory mechanisms that exclude women or generally rule against them.

This tendency is aggravated by land scarcity, the State's weak capacity for enforcing inheritance rights in the face of commercial pressures on land, land restructuring programmes, urbanization and HIV/AIDS – which have increased the individualization of land rights in many countries, reducing the pull of traditional obligations and responsibilities, as described in Box 2.4 (Kapur, 2011; Harrington and Choprab, 2010). Specific strategies and mechanisms may therefore be needed to support the land inheritance rights of widows, including practical support, legislative reviews and land registration, as the examples in Box 2.5 suggest.

BOX 2.4

Women's tenure security and HIV/AIDS

Source: Wehrmanm, 2003.

HIV and AIDS can lead to increased tenure insecurity for widows and orphans, who suffer more incidences of property grabbing when their husbands and fathers die of AIDS. Young widows are particularly vulnerable because they have not had time to develop strong social networks in the vicinity of their marital homes, are less likely to enlist the support of local leaders, and do not have children who are old enough to resist on their mothers' (and their own) behalf.

Elements of the government's land administration system can help to mitigate these impacts of HIV/AIDS on land tenure. For example, land boards can play a valuable role in preventing land sales that would compromise the welfare of widows and orphans. It is therefore important that women have equal representation and voice on these boards and that all board members are sensitive to the interests of vulnerable people in the community. Local-level dispute resolution/mediation mechanisms can also play a vital role in protecting vulnerable people.

It is important that women obtain joint title or titles in their own names, to ensure their tenure security. Civil registries should be linked to land registries, and consent requirements should be tightened to prevent abuse of the inheritance rights of widows and orphans. Registrars at civil and land registries should be trained on these issues, and registration forms should be specifically designed to enable joint registration of land and property. One of the greatest needs is awareness raising and legal assistance for vulnerable groups and individuals, who often do not know their rights or how the land administration system functions.

BOX 2.5
Good practices for dealing with inheritance issues

Sources: YWAP, 2011; Rebecca Sittie, personal communication.

In *Kenya*, the Young Widows Advancement Program carried out action research funded by the International Land Coalition to assist young widows who live in Nairobi and were married in Nyanza Province. The project helps the widows to avoid dispossession of their property and to reclaim any land rights lost on widowhood. The Kenyan Succession Act of 1981 stipulates that women may inherit property as dependants of their deceased husbands. However, young widows frequently experience property grabbing because they do not have adult sons to defend them and are less established in their husbands' home communities and families. In the local Luo culture, the writing of a will is seen as a bad omen signalling death; the prevalence of HIV/AIDS therefore discourages husbands from writing wills protecting the rights of their widows. Cultural difficulties also make it difficult for young widows to assert their inheritance rights. Some male chiefs are wary of accusations of inappropriate relationships with the young widows if they help them to assert their land inheritance rights. Overall, however, the project has been successful in empowering young widows through easier access to justice and the information they need to take forward their claims.

In *Ghana* too, the law provides for equal ownership of property in civil marriage. When land is jointly registered, it can be harder for a deceased husband's family to chase his widow off her land or out of her matrimonial home; she already owns 50 percent of the property and, by law, inherits some of her deceased husband's 50 percent, making it easier for her to buy her in-laws' share. In recent years, men have supported joint registration to ensure this kind of tenure security for their wives.

Tenants and their families also have concerns over land inheritance and succession. In some countries, entire families may be evicted from the land they use or occupy when the formal tenant dies. Often, only male household heads are named on land leases. Surviving family members risk eviction unless tenancy rights are protected by legislation on inheritance (FAO, 2001: 29–34).

CHECKLIST 2.1:
Supporting gender-equitable inheritance

Sources: WOLREC, 2011; YWAP, 2011; RWN, 2011; FAO, 2001: 29–34.

✓	Make sure that the policy and legal framework explicitly recognizes the inheritance rights of widows (and widowers) and that effective enforcement and monitoring mechanisms are in place.
✓	Ensure that professionals in the land, family and justice sectors are aware of how land laws intersect with family laws.
✓	Organize public sensitization and awareness-raising sessions on existing laws, so that people – especially women – know about the laws that protect them.
✓	Train customary authorities on statutory inheritance laws.
✓	Promote community discussions on: marriage and inheritance practices, friction between these practices and statutory law, and possible solutions for aligning the two.
✓	Where statutory protection is inadequate or inaccessible, help women to use customary laws to negotiate their land rights, for example by asserting the rights of widows to secure land because of the bride-wealth their families paid.
✓	Develop the capacity of organizations supporting women's rights and recognize their role in raising awareness and undertaking strategic interventions to support widows.
✓	Support legal empowerment through better access to justice, dispute resolution mechanisms and legal support that enables widows to claim their rights.
✓	When legal protection applies only to civil marriages, sensitize youth on the value of marrying legally and ensuring that all children are born in wedlock.
✓	Encourage will-writing where local culture supports this.
✓	Name both spouses as co-tenants on land lease contracts, so that the tenancy remains legally valid should one of them die.

REMEMBER!
Responsible governance of land tenure – governing land for women and men – requires understanding of the complex links between land and family laws and support to gender-equitable inheritance.

Compulsory land acquisition can be problematic in a context of legal pluralism. Issues include deciding who should receive the compensation for jointly owned land, and resolving sibling or inter-generational conflict over who owns the land. It is good practice for laws to require identification of all the people and entities that suffer losses from compulsory land acquisition, and to create compensation mechanisms that ensure joint family decision-making about use of the compensation received (FAO, 2008a: 33–34). The same applies in cases where lands, particularly in developing countries, are bought or leased by transnational, commercial actors.

The multiple land tenure arrangements and institutions of legal pluralism can provide vulnerable people with several options for securing their land rights, allowing them to fall back on customary systems when the formal system is ineffective or inaccessible (Mackenzie, 1989; 1990; 1993). Although this is not a long-term solution, it is important to facilitate all women and men – and not just the powerful – in realizing and exercising their land rights under whichever formal or customary system best accommodates their interests. This requires actions such as public awareness raising, gender sensitization within statutory and customary land tenure governance institutions, and the gradual establishment of participatory, transparent and accountable legal mechanisms and processes.

For gender-equitable governance of land tenure, the existence of legal pluralism therefore implies that attention must go beyond policies, laws and participation issues to address underlying perceptions of gender relations and power structures. This requires a very long-term perspective and patience, in addition to communication and awareness raising, as explored in *Module 5*.

CHECKLIST 2.2: **Supporting gender equity within legal pluralism**

Sources: WOLREC, 2011; Nakayi, 2010.

✓	Work constructively with customary and religious institutions and their leaders, and involve the whole community in decision-making processes and outcomes involving customary law and land tenure practices.
✓	Raise awareness on how things can change from the customary perspective. Support women's self-help groups as these can play a central role in supporting cultural change.
✓	Ensure that the recognition of customary laws in parallel with formal law does not lead to the manipulation of customary laws in ways that are detrimental to vulnerable people.
✓	Establish two-directional training programmes, with customary/religious leaders training formal judges on customary/religious law, and the judges training them on statutory law. This helps build bridges between different legal systems and can improve interactions to ensure overall outcomes that are more just.
✓	Find out where women and men obtain assistance and legal protection. Support these mechanisms and processes and facilitate people's ability to benefit from legal pluralism.

REMEMBER!
Written law is the best protection against gender-inequitable customary or religious practices. However, there are wide gaps between law and practice, and bad practices continue unchecked.

Drafting of laws, regulations and procedures

Good land laws often remain unapplied because of the lack of secondary legislation and regulations and procedures for their implementation. It is therefore essential to improve land law enforcement and develop gender-equitable regulations, procedures and strategies for implementation (FAO, 2008a: 12). Table 2.1 describes some of the legal instruments that have to be drafted and enforced as part of an implementation strategy for gender-equitable governance of land tenure.

TABLE 2.1:
Legal instruments

LEGAL INSTRUMENT	DEFINITION
International conventions	Treaties, conventions or other forms of legally binding international instrument, which confer legal obligations on the countries that have ratified them.
The Constitution	The document that legally establishes the basic principles by which a country is governed. Many constitutions are written, some are unwritten and are established over time through precedent and case law.
Primary legislation	Laws made by the legislature (the legislative branch of State power). Unless they are very detailed, primary laws may need supporting secondary legislation.
Secondary legislation	The subordinate regulations, rules, orders and decrees required to set out the detail of primary laws, to ensure that they are applied objectively and fairly.
Rules and procedures	The documents that guide officials and administrators on how laws should be implemented.

Primary legislation must comply with the Constitution, and secondary legislation, rules and procedures must comply with both primary laws and the Constitution. If the Constitution includes protection for fundamental rights, by upholding principles such as gender equity, for example, then these principles must be observed both by the legislature making primary laws and by the bodies responsible for adopting subordinate laws.

Just as the land policy-making process is different from the content of the ensuing land policies, so is the process of drafting legislation different from the content of the resulting laws, regulations and procedures. A gender-equitable drafting process does not necessarily lead to gender-equitable legislation. However, gender-equitable participation in the drafting process is more likely to result in land laws and subordinate laws that respond to the different needs of women and men from different social groups, and that therefore enable gender-equitable implementation of the overall land policy and laws.

Box 2.6 describes measures for ensuring that legal drafting processes are participatory, consultative and inclusive for both women and men.

BOX 2.6
Consultative, participatory and gender-equitable legal drafting

Sources:
FAO 2002c: 31–32; Government of Samoa 2008.

In *Mozambique*, drafting of the 1997 Land Law and its accompanying regulations established an important precedent for involving civil society, government and other stakeholders in developing policy and changing national land legislation, and is widely cited as a good practice in innovative land law reform. The 1996 National Land Conference that preceded the Land Law's presentation to the National Assembly was attended by more than 200 people from a wide range of constituencies, including national and local NGOs, small organizations of farmers and women from different areas, the private sector and independent land sector professionals and experts. The draft law was debated point by point over the three-day conference, and received broad support.

In *Samoa*, the *Legislative Drafting Handbook* of 2008 identifies steps that government ministries and agencies must undertake before embarking on legal reform and drafting legislation, including regulations and procedures for implementing existing laws. For example, the instructing ministry must consult relevant stakeholders as soon as a law/regulation has been drafted, and the stakeholders' comments must be incorporated in subsequent drafts. Different groups of people are consulted to obtain the views of a representative cross-section of the entire community – women, men, youth and the elderly – and all consultation outcomes are recorded. The handbook also stipulates that all laws and regulations must use gender-neutral terminology; in 2007, when all Samoan laws were consolidated, masculine terms were replaced with gender-neutral ones.

All programmes for land law reform – including those funded by donors – should support the drafting of subordinate laws, regulations and procedures, such as operational and procedural manuals for land registration. These supporting instruments can make or break a law, so designing and disseminating regulations, procedures and manuals that promote an overall gender-equitable approach to law design and implementation is key to ensuring gender-equitable tenure governance (FAO, 2011a: 13).

In addition, the institutions involved in land tenure governance must be committed to ensuring that regulations are drafted and implemented properly. For example, the documentation and procedures required to support land registration claims must be available and accessible to women as well as men; this may require revising the relevant regulations to provide women and men with equal opportunities to register land (World Bank/FAO/IFAD, 2009a: 145). Regulations should also enforce gender-equitable governance of land tenure more broadly, such as by requiring joint titling for the jointly acquired land of married couples, and proof of consent from both spouses for the formal recording of transactions of marital property (World Bank/FAO/IFAD, 2009a: 145–146).

Regulations about agricultural leasing arrangements should also take gender equity into account. Leasing arrangements are the only or main source of access to land for many farming families across the world, and are therefore an important part of land tenure. Good practices in support of gender-equitable leasing arrangements include (FAO, 2001; 2004):

- allowing both women and men to be named as leaseholders; allowing family members to lease land independently of the household head; and including on formal leases the names of all the family members who use or occupy the land, along with the name of the leaseholder –the use of model leases can be particularly beneficial here;
- ensuring that the landowner named on a lease has the legal capacity to lease the land – for example, the landowner may have customary law obligations to seek the consent of other family members;
- on the death of a tenant, enabling the surviving spouse and family to remain on the land until the end of the tenancy or until more permanent arrangements can be made, retaining all the rights available to the tenant named on the lease.

All land laws need strategies and guidelines for implementing their gender-sensitive and gender-equitable provisions, and for ensuring effective, responsible and gender-equitable implementation of the law (FAO, 2011a: 13). Legal drafting guidelines such as those in Samoa (Box 2.6) are particularly useful in this area. Establishing standardized legal drafting procedures helps to ensure that details such as the design of forms and appeal mechanisms protect the rights of both women and men.

The piloting of regulations or laws tests how well the prescribed procedures work in the field (Rachael Knight, personal communication) and can be particularly useful where legislation has been drafted with little attention to the institutions and processes needed for its implementation. After piloting, feedback on how effective the new laws or regulations have been can guide amendments and improvements. Piloting may be most useful for testing the gender equity of operational and procedures manuals and fine-tuning the details of existing laws.

It is also important to establish a higher body with responsibility for monitoring gender equity in the implementation of land laws and regulations. For instance, in Rwanda, the Gender Monitoring Office monitors compliance with the indicators of its gender equality programme. Such an office is likely to be more effective -in supporting compliance – particularly with rules such as quotas for women's participation in governance institutions, committees meetings, etc. – than appeal mechanisms involving courts, special land courts or tribunals, which tend to be reactive, slow, expensive to appellants and uncertain of success (McAuslan, 2010: 126).

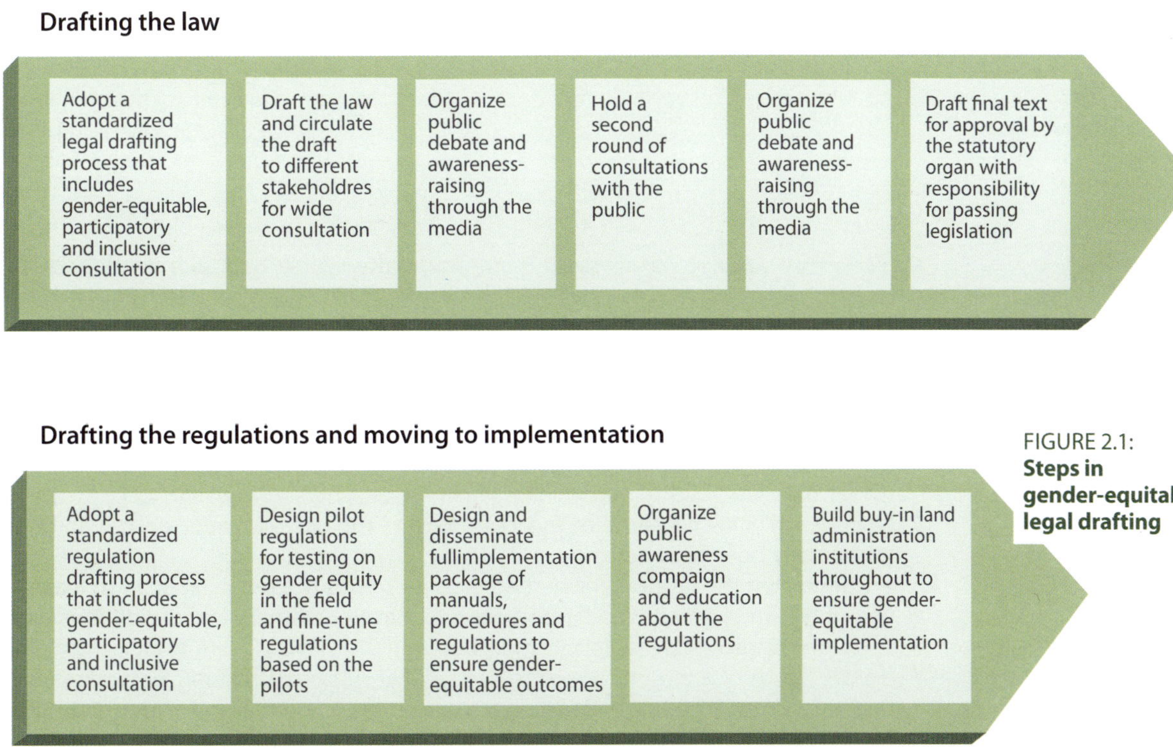

FIGURE 2.1:
Steps in gender-equitable legal drafting

BOX 2.7
Lessons from local legal drafting

Source: Knight, 2011

The International Development Law Organization conducted a pilot project on community land titling in *Liberia, Mozambique* and *Uganda* between 2009 and 2011. Working with national and local civil society partners in each country, the project's aims included understanding how to support communities' efforts to document land claims and devising and piloting strategies for facilitating gender equity and preventing intra-community injustice and discrimination.

An important component of the project was a participatory process for establishing rules for community land administration (by-laws/community constitutions). Meetings were held at each of the main steps in drafting by-laws: brainstorming and recording existing rules; eliminating or modifying rules that no longer suited the community's needs and adding new ones that did; checking consistency and compliance with the national Constitution and national or international laws; and, after a final review, preparing a final draft for formal adoption as the community's by-laws or constitution.

From the outset, the project teams stressed that the whole community had to be involved throughout the process and that men as well as women representatives of households had to attend the final meeting to endorse the by-laws.

This local-level legal drafting process enabled communities to reflect on and discuss their existing land tenure governance rules and the underlying reasons for them, including the traditional discriminatory rules that women questioned and challenged in the meetings. Women started to see that they could argue for greater land rights within their communities, and communities were encouraged to strengthen the land rights of vulnerable groups and the intra-community mechanisms for safeguarding existing rights. The project also enabled the alignment of local rules with statutory law, increasing community and women's awareness of both national law and human rights. Through women's involvement in the meetings and discussions, community perceptions of women's place in the governance of land tenure began to change, and women's equitable participation in the local legal drafting process influenced the content of the ensuing community rules.

Community constitutions and by-laws should be written up or drafted at the local level in a way which allows the gender-equitable participation of all the women and men in the community and can help boost local people's legal literacy and understanding of statutory law. Local legal drafting processes can open up spaces for women to question the rules that disadvantage or discriminate against them and to advocate for rules that strengthen their land tenure security and role in local land tenure governance (Knight, 2011). Box 2.7 describes how gender equity was successfully achieved in community land titling processes in Liberia, Mozambique and Uganda.

CHECKLIST 2.3:
Supporting gender-equitable legal drafting

Source: McAuslan, 2010.

- ✓ Make sure that all land legislation and regulations are consistent with the overarching constitutional framework and international human rights law and that this framework is supportive of gender equity.
- ✓ Define processes for legal drafting that include gender-equitable consultations and participation, ensuring that women's groups are involved and their contributions taken seriously into account.
- ✓ Hold women-only, men-only and joint consultations, allowing the emergence of diverse views and needs. Women and men from all social groups should be represented in consultations, and constraints to participation must be taken into account.
- ✓ Set gender quotas for legal drafting committees, to support the increased participation of women.
- ✓ Adopt measures to ensure that laws and regulations cannot be passed without women being consulted.
- ✓ Design operational and procedural manuals for land administration and management that describe how to implement the law effectively in support of gender-equitable land tenure governance. Pilot the regulations and adjust them accordingly.
- ✓ When setting quotas for women members of local land committees and commissions, do not assume that women are always willing and available to sit on such bodies, especially in rural areas; support women's participation by ensuring that meetings are held at times that are generally convenient for women.

REMEMBER!
Effective gender-equitable strategies and processes for implementing land laws are essential for responsible gender-equitable governance of land tenure. Legal and regulatory drafting that is gender-equitable in both method and substance will support this from the start.

Gender-equitable land tenure governance and exercising rights to land

Establishing a solid legal foundation for responsible gender-equitable governance of land tenure requires paying attention to the different elements involved in enabling women and men to realize and exercise their rights to land. These elements include access to justice, dispute resolution mechanisms and legal support, and legal literacy, as defined in Table 2.2. The first three are discussed in the remainder of this module, while legal literacy is discussed alongside the closely linked issues of communication and awareness raising in *Module 5*.

TABLE 2.2: **Elements in exercising land rights**

ELEMENT	DEFINITION
Access to justice	The ability to pursue land claims with recourse to relevant legal and judicial systems and customary or statutory dispute resolution mechanisms. Under international human rights law, access to justice also includes the right to an effective remedy for anyone whose right has been proven to be violated.
Dispute resolution mechanisms	The statutory or customary processes and institutions used to resolve land disputes within the overall justice system
Legal support	The practical support needed to pursue individual land claims: information on procedures, financial assistance for related costs (legal aid), and professional support from paralegals or lawyers (legal assistance)
Legal literacy	Knowledge of the law and how to implement/enforce it, and understanding of the relationship between law and practice

Access to justice

Access to justice varies among different groups of people in every country context. For example, the legal support needs of women in rural areas will differ from those of women in urban areas, as will those of literate women and men from those of illiterate women and men, and those of women who rely on customary law for their land rights from those of women who have acquired and/or registered land formally in their own names (Nakayi, 2010). Responsible gender-equitable land tenure governance should therefore involve carefully targeted approaches to making access to justice on land issues gender-equitable for all groups of women and men.

However, in many countries, access to land justice for women is limited by cultural, social and institutional barriers. These can include high court fees and related costs, lack of knowledge about land rights and the formal justice system, physical distance, language barriers, under-representation in the judiciary, lack of confidence and autonomy to act independently in legal proceedings, and socio-cultural norms that make women fearful of retribution or ostracism if they pursue land claims. As a result, women tend to be denied access to justice more often than men, and are also more likely to be denied justice altogether (UN Women, 2011: 54; FAO, 2006a: 154–156; GROOTS Kenya, 2011).

Under legal pluralism, particularly in rural areas of developing countries where State land justice institutions may be absent or inaccessible, customary law is often the primary means of resolving land disputes and enforcing community rules (Knight, 2011: 148). Women may find it easier to approach customary authorities than formal justice institutions, if the former are more local and closer to home. However, customary law and authorities may not protect women's rights as well as the formal justice system would because of gender biases and male dominance.

UN-Women defines the justice chain as "the series of steps that must be taken to access justice" (UN-Women, 2011: 48). A gender-responsible justice chain requires (UN-Women, 2011: 56–57 and 63):

- clear mandates and procedures, including standardized protocols, regulations for enforcing laws, and coordination among different parts of the justice system, to ensure that public services are responsive to women –adequate and sustained funding, and measures for monitoring implementation are needed;
- integrated and specialized services to help women overcome the institutional and social barriers to land justice that they face. One promising approach is integrating services, such as at one-stop land offices where all land services are under one roof, reducing the number of steps needed to obtain access to justice; it may be helpful to have women working at the front desk of land offices, as they may be more accessible and approachable for women clients, as discussed in *Module 4*;
- support for women's participation in the justice sector, to increase gender-balanced representation across all aspects of governance and to boost public confidence in the justice system as a whole;
- sensitization and training to make judges aware of the barriers that women face in obtaining access to land justice, so that attitudes change. Simple but effective solutions – such as waiving court fees, providing forms free of charge or prioritizing sensitive cases – can be implemented.

CHECKLIST 2.4:
Increasing gender equity in access to land justice

Sources: Nakayi, 2010; UN Women, 2011.

✓ Provide targeted information and training for women and men on access to land justice.

✓ Address the underlying social and institutional causes of inaccessibility, such as culture, poverty, illiteracy and distance.

✓ Support women's increased representation and participation in land administration and management institutions, especially in land dispute resolution mechanisms.

✓ Recognize the role of customary law and authorities in land dispute resolution processes and their functioning. Support customary authorities in becoming more gender-equitable, and identify the reasons why customary practices exist and how they could support gender-equitable governance of land tenure.

✓ Encourage the transformation of customary law in ways that also meet the interests of powerful men, to minimize any temptation for them to abuse their power and create new barriers to access to land justice. Specifically, when working or advocating with men working in the justice sector adopt a strategy of dialogue rather than confrontation.

REMEMBER!
Barriers to access to land justice affect all vulnerable people, although women have tended to be particularly badly affected. Responsible gender-equitable governance of land tenure requires access to justice for all women and men across castes, religions, ethnicities and ages, to enable them to realize and exercise their rights to land.

Land dispute resolution

Supporting women's increased representation and participation in formal and statutory land dispute resolution mechanisms is an acknowledged means of improving women's recourse to land dispute resolution and justice (World Bank/FAO/IFAD, 2009a: 24). Box 2.8 illustrates this with reference to Rwanda, where increased women's participation in local dispute resolution committees made gender-equitable outcomes more likely.

BOX 2.8:
Gender-equitable land dispute resolution

Sources: RWN, 2011; Daley, Dore-Weeks and Umuhoza, 2010; International Justice Mission, no date.

The *Rwanda* Women Network (RWN) undertook a study in Bugesera District, funded by the International Land Coalition, to assess women's experiences in the implementation of laws that protect their land rights. The study describes the importance of the abunzi (local mediators) in protecting women's land rights, especially in inheritance-related and other intra-family disputes, which were the most common form of dispute involving women. The study found that most women chose to report their disputes to local authorities, while some went directly to the abunzi. Only a few women reported their disputes to their family councils, which were considered less trustworthy than either local authorities or mediators. At least 30 percent of the membership of local authorities and abunzi must be women, which helps explain why women with land cases to report are more likely to trust them.

The study highlights the importance of local-level alternative dispute resolution processes and institutions, particularly in countries such as Rwanda that have suffered serious conflict and whose existing customary institutions may not be trusted to be objective. The abunzi themselves called for:

- encouragement of the rule of law and regular training on land laws for local dispute resolution committees;
- broader gender sensitization and awareness raising on land laws for the whole population, including special training on gender issues for men who retain discriminatory beliefs and values – one male respondent in the research pointed out that "As men, we should also value women as we value ourselves".

Abunzi training programmes have been carried out by government and civil society, including by the NGO International Justice Mission (IJM) in cooperation with the Ministry of Justice. Lawyers working with IJM provide practical legal support to victims of illegal property dispossession, especially widows and orphans, to help them obtain legal redress in Zambia, Uganda and Kenya as well as Rwanda, and work in all four countries to sensitize local leaders and members of local dispute resolution committees, both men and women, on relevant issues.

In Kyrgyzstan, from 2003 to 2010, UN Women's Land in the Right Hands programme helped develop the knowledge and capabilities of major actors in the justice sector in employing a rights-based approach to gender equality and women's empowerment. Specific attention was paid to strengthening partnerships between legal offices and experts from the Department of Cadastre and Registration of Immovable Property, resulting in the prompt settlement of disputes in the courts and restoring women's legal rights to land shares and the holding of title documents (Djusaeva, 2012).

One reason why women have particular difficulty in ensuring recognition of their land claims is that they often have very little power and authority to take public action. Land disputes involving claims by women are often considered private, household matters requiring resolution at that level. Women, particularly in rural areas, often find that neither the formal nor the customary tenure system recognizes their rights to land (FAO, 2006b: 48).

Alternative dispute resolution is an increasingly popular solution. Alternatives include consideration of court-based mediators to help people who have lodged court proceedings to resolve their disputes without having to wait for the court process (Magezi, 2010).

In many countries, customary authorities and institutions are still largely responsible for local land dispute resolution. Customary dispute resolution mechanisms may be speedier and more accessible and have greater social legitimacy than statutory ones, but they are often male-dominated and gender-biased. Reforming them to ensure gender equity may enable them to work more effectively while retaining their comparative advantage (FAO, 2006a: 158 and 166).

The challenge is to work with customary institutions to support gender-equitable governance of land tenure while seizing opportunities from decentralization processes in land tenure governance to make formal statutory dispute resolution mechanisms more accessible to rural women and men. Box 2.9 provides some good practice examples of working with customary institutions in land dispute resolution; *Module 3* addresses broader issues of working with customary institutions.

Land disputes can also be resolved within the proceedings of government justice institutions, such as the Ugandan Department of the Administrator General, which is responsible for managing and distributing the property of deceased people in a number of different situations, including when they have died without making a will (Namutebi, 2010). Such institutions can be important when existing inheritance laws are discriminatory but new laws have not yet been put in place. The Ugandan Department of the Administrator General's efforts to ensure the participation of women, especially widows, have helped to ensure more gender-equitable outcomes in upholding the distribution of property to women while avoiding court action (Namutebi, 2010; Ssonko; 2010).

BOX 2.9:
Working with customary institutions in land dispute resolution

Sources: Harrington and Choprab, 2010: 20; Carfield, 2011; ULA, 2010; Ahikire, 2010.

In *Kenya*, among communities in Nyanza province, in-laws were routinely seizing lands from widows and pushing them into dire poverty, while community elders proudly proclaimed that their culture protected women. A Kenya National Commission for Human Rights programme provided the space for widows to tell these elders their stories of being chased from their lands. The elders, caught between securing patrilineal land rights and protecting women, had to develop innovative means of resolving this dilemma. In many cases, they began to defend the widows and ensure that they received a life interest in their deceased husbands' land. This was for the widows' own protection, but also to ensure that their sons would subsequently receive a portion, as dictated by patrilineal inheritance practice.

In war-torn northern and eastern *Uganda*, internally displaced persons have begun to return home and re-establish claims to their ancestral land. This has caused many conflicts over competing claims and encroachments between neighbours and families driven by increasing land scarcity. In 2009, the Uganda Land Alliance launched a pilot project aiming to improve access to land justice through the empowerment of traditional authorities and the integration of the customary and formal justice sectors. A project assessment in 2010 found that when they had land disputes, women and poorer people generally approached customary leaders first. The project therefore worked to improve the existing customary justice system by developing the technical capacity of traditional authorities to address land rights violations. This involved both training for traditional authorities and sensitization for communities. Land rights centres were set up to train paralegals and establish a new system for mediating land disputes, in which paralegals work with local political leaders and traditional authorities to apply the traditional justice approach of seeking mutual understanding rather than the "zero-sum game" approach of the courts.

Throughout *Uganda* there is evidence of women utilizing both customary and Local Committee (LC) courts, although most women stop at lower levels (clan, LC1 and LC2 courts) and only those who can afford it, go on to LC3 and magistrates' courts or district land boards. The LC courts are accessible, do not require representation by a lawyer, use familiar language and accept many different forms of evidence. However, except for LC3 courts, they are not always seen as impartial and they may be "too close" to their users. Nonetheless, judgments from LC courts, which utilize a blend of knowledge, formal statutory law, custom and norms, have the potential to recreate and document positive customary norms and practices in support of gender equity.

Legal support

Efforts to increase gender equity in access to land justice should consider all the dispute resolution mechanisms and justice systems available and should facilitate the access of women and men to those that are most appropriate to their needs. Support should start by intervening at the top, to ensure that governments establish relevant measures, including legal aid and practical legal assistance. In Tajikistan, women's access to free legal services has been improved through the establishment of district task forces with staff from women's committees and hired lawyers (Undeland, 2012). Box 2.10 describes another potential good practice from Samoa.

In some countries, legal aid does not cover cases relating to land, inheritance and family law. In addition, not all legal aid schemes take gender issues into account. Exceptions include India, where women are specifically listed as beneficiaries of the Legal Services Authorities Act of 1992, and Ghana, where the Legal Aid Scheme Act of 1997 covers not only criminal matters but also civil matters involving inheritance rights (FAO, 2006a: 160).

It is good practice for land administration projects to review their procedures for land dispute resolution and legal support to ensure that both men and women have access to relevant services and are treated equitably. This might require special measures to facilitate women's access to project staff and government offices, such as by bringing land dispute resolution mechanisms to the local level or providing legal support through the project, as the Kyrgyzstan case described in Box 2.11 has done.

BOX 2.10:
Legal support for customary landowners

Source: Josephine Stowers-Fiu's presentation on "Capacity building and gender equity in land administration institutions in Samoa" at the May 2011 technical workshop.

In *Samoa*, more than 80 percent of land is held under customary tenure arrangements, which provide equal rights to control and manage customary land for women and men, under bilineal succession and informal leasing. However, ultimate ownership and authority over land rest in the extended family, personified by a custodian/trustee known as the Matai, who is chosen by family consensus and controls and directs the use of land and family assets, including the leasing of customary land to others. Decisions made by Matai are sometimes contested in the Lands and Titles Court. Although most Matai are men, there are a few women Matai.

The Customary Land Act of 1965 provides for formal leases of customary land between the Minister of Natural Resources and Environment (on behalf of the customary landowners) and the users of the land (the lessees). The ministry administers the lease application process on behalf of the customary landowners, to avoid the conflict of interest that arose in the past, when customary owners relied on the lessees' lawyers to prepare the documentation. The ministry also informs the owners about the leasing process, including by explaining all the terms and conditions of the lease, to ensure that they reflect the wishes of all the customary landowners (the extended family).

Although men and women have equal rights to access to land, many women lack awareness of their rights. Unless the Matai is a woman, women are often not involved in the negotiation and discussion of leasing arrangements. Having more female staff advising customary landowners about leasing arrangements would strengthen the benefits of the legal assistance provided by the government, by making the leasing process more inclusive, accessible and gender-equitable.

BOX 2.11:
Legal assistance to rural populations

The Legal Assistance to Rural Citizen's Project (LARC) in *Kyrgyzstan* – implemented between 2003 and 2006 with support from the World Bank, the United Nations Development Programme, Switzerland and the United States of America – offered legal services and aimed to teach rural people how to apply the law to resolve their land disputes. Women and men applied for project assistance and sought help with similar disputes, involving village or local authorities attempting to take the land the claimant had received during national land reforms and assign it to other men in the village. The project recorded no land disputes arising from divorce or inheritance, suggesting either that this type of dispute was uncommon or that women were reluctant to take cases against relatives to court.

Source: World Bank/FAO/IFAD, 2009a: 150–151.

In one intra-family case, a disabled woman had been given the family house by her father, with the appropriate official documentation. However, the woman's uncle and cousins refused to leave the house and drew up an alternative document, certifying their right to it. After the local authorities discouraged her from taking the case to court, and having failed to advance her case with the local courts, the woman approached LARC and eventually reached consensus with her extended family. The LARC project thus assisted women with land disputes, but its experience suggests that strong legal support and assistance are needed to improve gender equity in land dispute resolution where cultural norms discourage women from pursuing cases against relatives.

Civil society organizations have a particularly important role in providing legal support for gender-equitable governance of land tenure. Legal organizations can offer practical legal assistance in individual cases, while also often having the capacity and mandate to pursue strategic litigation and take up important test cases in the courts (UN-Women, 2011: 118; World Bank/FAO/IFAD, 2009a: 151; FAO, 2006a: 160; Kapur, 2011).

Test cases can help to enforce the implementation of gender-equitable provisions in land laws and land-related regulations, and ensure gender-equitable governance of land tenure. For instance, where customary law is widely applied, test cases can involve reform of the more discriminatory aspects of customary law, rather than its complete replacement (FAO, 2006a: 166). Legal support from civil society can therefore make courts a useful vehicle for enforcing gender equity.

Civil society organizations are often most effective when they work in cooperation with governments, as described in Box 2.12 with respect to Pakistan, where civil society support was critical in improving the gender equity of tenure governance under the Land Distribution Programme.

BOX 2.12:
Good practice in legal support

Source: ILC, 2010c.

A project in *Pakistan* – funded by the International Land Coalition in 2009 and 2010 – established legal aid committees (LACs), with sub-committees of lawyers, in five districts of Sindh Province. The LACs took up the cases of women facing appeals against their allotments of land under the Government of Sindh's Land Distribution Programme (LDP), and deterred the former landowners from appealing against the granting of their land to women beneficiaries.

Women's organizations (WOs) were formed in the five districts, and the project trained WO members in organizational management and in providing assistance to women with problems in attaining their legal land entitlements under the LDP. The LACs and WOs collaborated to provide legal services to 50 women whose recently granted land was under litigation; half of these cases were successfully solved. Another 36 cases were withdrawn by the former landowners, following advocacy by the LACs and WOs.

The project learned that providing legal assistance requires a long-term commitment from donors and funders, as some of the cases supported by the LACs were referred higher up the formal justice system; the women concerned will require continuing legal aid, and the LACs and WOs will need to develop their own capacity to provide this support at higher levels of formal justice. The project also learned that in addition to holding governments accountable for gender equity in their land-related programmes, it is also important to have a strategy for engaging directly with officials working in these programmes.

Civil society and international organizations can also train community paralegals to work on gender issues at the community level, as illustrated by the examples from India, Mozambique and the Commonwealth Secretariat in Box 2.13. Community paralegals can provide relatively low-cost legal support as they require less training than fully qualified lawyers. They are also important in efforts to boost legal literacy, as discussed further in *Module 5*.

BOX 2.13:
Training community paralegals for work on gender issues

In *India*, the South Asia Rural Reconstruction Association (SARRA) provided community paralegal training for forest dwelling tribal people in Andhra Pradesh State in 2010, in collaboration with senior government officials responsible for forests, environment and tribal issues. A four-day paralegal training workshop for 63 tribal people (mainly women and youth) included direct interaction with a state minister and other senior officials. This was the first time such direct interaction had taken place and it boosted the tribal peoples' confidence, as well as enhancing dialogue and improving tribal people's knowledge of the legal and administrative procedures for obtaining land titles. SARRA translated the 2006 Land Rights Act into the local language (Telugu) and printed 1000 copies, along with practical information for tribal people on how to apply for land assignments. Almost half of the newly trained paralegals completed their own application forms for land assignments, and assisted 45 other tribal families in doing the same.

Under its three-year Gender and Land Rights project in *Mozambique*, with support from Norway, FAO and its government institution, Centre for Juridical and Judicial Training (CFJJ) have started paralegal training courses and field support on gender for civil society, the public sector and community leaders. A broader training programme on land for judges, lawyers and police officers, in operation since 2000, demonstrated the need to disseminate information about the new land laws to local government and civil society actors working directly with rural populations. Paralegal training courses started in 2005, and the project developed a gender module in 2010, seeking to work on gender issues with men as well as women, through paralegal courses, district seminars for judges, prosecutors and police chiefs, a field technical support programme, and community sessions. The project is also raising awareness on women's land rights and gender equality through a national advocacy campaign. Messages are disseminated widely in all four main national languages, including on T-shirts, banners and calendars, in comic books and videos, and through community theatre, radio and other media. The project is producing a textbook for incorporation into the course materials of the broader paralegal and justice sector training programme.

A *Commonwealth Secretariat* handbook, launched in July 2011 at the Law Ministers' meeting in Australia and disseminated in partnership with ActionAid, aims to "provide substantive and practical information to all relevant actors working to effect women's claims to resources and rights under the law and custom". The handbook is a practical tool for magistrates, local administrators, traditional chiefs, community leaders and rural women at the grassroots level, and will be particularly useful as a training and reference guide for community paralegals. Published in several languages, it provides step-by-step guidance for developing a comprehensive understanding of the law and encouragingthe law's full application in protecting women's access and rights to land.

Sources: Catherine Gatundu's presentation on "Africa experiences of civil society involvement in land policy process" at the May 2011 technical workshop; Marianna Bicchieri's presentation on "Gender and Land Project – Mozambique" at FAO, Rome, 13 May 2010; ILC, 2010d; Catherine Gatundu, personal communication.

The project discussed in Box 2.7 also employed paralegal support as a way of supporting community land titling processes. This project was particularly innovative in seeking to establish the most effective levels and types of legal support for ensuring gender-equitable land tenure governance in local communities, as described further in Box 2.14.

BOX 2.14:
Lessons on legal support

In the project on community land titling in *Liberia*, *Mozambique* and *Uganda* described in Box 2.7, groups of communities were provided with different levels of legal assistance while they followed their nations' legally mandated processes of community land titling. The groups' progress was monitored, and the results were compared and analysed to identify how international organizations, governments and national NGOs might best support the implementation of community land titling legislation.

In each country, 20 communities were requested to participate in the project; each community was assigned to one of four groups receiving different levels of legal support:

- monthly legal education and training provided by a field team of a lawyer and a community mobilizer/technician – all community members were invited to take part in these training sessions, and specific measures were adopted to ensure the participation of women;

Sources: Rachael Knight's presentation on "Gender-equitable participation in land governance: community land titling in Liberia, Mozambique and Uganda" at the May 2011 technical workshop; Knight 2011; Catherine Gatundu, personal communication.

- paralegal support by two community-based and elected "land paralegals" (one woman and one man in Mozambique and Uganda, at the community's choice in Liberia) plus monthly legal education and training – the paralegals received in-depth training, including on the inclusion of vulnerable groups, and strategies for aligning customary rights with national laws and human rights principles;
- monthly legal education and full technical support from the project lawyer and field team throughout the community land titling process, including assistance with conflict resolution;
- control groups – communities attended one meeting where they were provided with copies of national land laws and regulations, a guide and other training materials on following the community land titling process.

Across all three countries, communities that received paralegal support and monthly legal education and training (second bullet point in the list above) had the most success in completing the registration process – outdoing those that received full legal support. Paralegals played a significant role in supporting community participation in the project. These findings suggest that training and supervising local elected community paralegals may be a low-cost, efficient and effective way of supporting large numbers of communities through land documentation processes. The project's experience also suggests that community land titling might best be supported by community paralegals supervised by a legal and technical support team.

A final important issue is the need to consider innovative solutions, particularly at the local level. Box 2.15 describes watchdog groups as a mechanism for legal support in Kenya. Watchdog groups can relieve the pressure on formal justice systems and serve as a vehicle for providing legal assistance to local people in local court proceedings and dispute resolution mechanisms. They can also double as a broader mechanism for participatory governance at the grassroots.

BOX 2.15: **Community watchdog groups in support of women's land rights**

Source: GROOTS Kenya, 2011.

The civil society organization, GROOTS Kenya, assisted grassroots communities in Gatundu District in forming community land and property watchdog groups (WDGs) to protect widows and orphans from losing their land and property.

The WDGs are based on the idea that community members – women and men – should work together to prevent property rights' violations within their community. Each WDG has 15 to 25 men and women members, who volunteer after a participatory community process. This process includes a community needs assessment; the mobilization of key stakeholders – such as village elders, human rights organizations and provincial authorities – through sharing assessment results and explaining the impact of land rights violations; and community dialogue. Women participants are particularly encouraged. When a land rights violation involves vulnerable community members, the WDG determines the facts of the case, alerts other members of the community, and mediates to ensure that the violator returns the property to the rightful owner. Various mediation methods are used, involving community leaders, local government officials, chiefs and elders. If necessary, the WDG arranges for the case to be filed in court and ensures that the court's ruling is implemented. The WDG also receives referrals from people or institutions such as members of the police, the judiciary, provincial authorities, district land registrars, land dispute tribunals and land control boards. Communities provide feedback through WDG-led evaluations and reviews, and successful WDGs are encouraged to share their progress and practices through peer exchanges or visits to other communities facing similar problems.

Through partners such as GROOTS Kenya and other agencies, the WDG members in Gatundu District have received basic training as paralegals, enabling them to provide mediation between disputing parties, and basic legal advice. The WDGs also build awareness and alert people to the proper channels for securing the documents that safeguard their rights.

WDGs in Gatundu have supported the provincial administration by dealing with cases at the grassroots level and reducing pressure on formal justice systems. Other innovations include WDGs' ability to make statements during land cases in the formal justice system. Through monthly reporting to the provincial administration on the performance of customary authorities, WDGs have proved to be an excellent means of participatory governance, providing valuable support for responsible gender-equitable governance of land tenure. Nevertheless, WDGs face challenges such as lack of recognition, financial and cultural constraints, lack of physical location (office space), and corruption.

CHECKLIST 2.5:
Ensuring gender-equitable legal support

Sources: WOLREC, 2011; YWAP, 2011; RWN, 2011; FAO, 2001: 29–34.

✓	Assess the type and level of legal support that communities require, including how best to facilitate protection of the land rights of women and vulnerable groups during community land titling efforts.
✓	Train community paralegals to work with both women and men, as men also need to support change. Provide community paralegals with financial, logistics and other support, particularly in supporting vulnerable groups such as widows, orphans and ethnic minorities.
✓	Set up community watchdog groups and connect them to legal aid centres for support.
✓	Set up mobile local courts and marriage, birth and death registration offices, to make it easier for people to obtain basic legal documents and to lodge cases in the formal court system when customary or local dispute resolution institutions cannot resolve them.
✓	Develop a database of court cases and legal precedents that uphold and support gender-equitable governance of land tenure at the national level, and develop awareness-raising campaigns to inform community watchdogs and paralegals about these examples.
✓	Encourage the documentation of good practices and learning from each other.

REMEMBER!
A varied and innovative approach to legal support will be the most effective. Governments should set an example from the top, but civil society organizations are also very important in reaching and supporting the grassroots.

Notes

[5] Marianna Bicchieri, Patricia Costas, Nigel Edmead, Debbie Fletcher, Renée Giovarelli, Rachael Knight, Martha Osorio, Rebecca Sittie, Josephine Stowers-Fiu and Margret Vidar provided substantial inputs to this module during the May 2011 technical workshop at FAO Headquarters, including through discussions of two legal working groups.

Summary of key messages on legal issues

It is important to have the right legal foundations for the legislative and policy framework to support gender-equitable governance of land tenure and for effective implementation and monitoring mechanisms.

Plural legal systems and customary law provide spaces for women and men to realize their land rights. Although they are not a long-term solution, efforts should be made to find these spaces and make them accessible to all.

Participatory and consultative drafting of laws, regulations and procedures, involving both women and men from different social groups provides essential support to gender-equitable governance of land tenure.

Access to justice is not always gender-equitable. It is necessary to work with both formal and customary justice systems, and both local and customary dispute resolution mechanisms to achieve gender equity.

Innovative approaches to legal support should be used, such as community watchdog groups and community paralegals. The level and type of support needed in each context should be carefully assessed.

Suggested monitoring indicators for legal issues

- existence of laws and polices supporting gender-equitable governance of land tenure
- % of women and men members of land law reform commissions
- % of women and men members of legal drafting teams working on land-related subsidiary laws, regulations and procedures
- presence of participatory and consultative legal drafting processes involving women and men from the grassroots
- % of women's groups' proposals taken into account and integrated in the adopted legal instruments
- existence of inheritance laws coherent with and supportive of tenure reforms and gender-equitable governance of land tenure
- number of enforcement and monitoring mechanisms for land laws
- % of women and men receiving training in access to land justice
- % of women and men members of land dispute resolution institutions and mechanisms, including local watchdog groups
- % of women and men members of customary authorities dealing with land
- number of initiatives for removing barriers to land justice for women
- % of women and men receiving legal support in land cases in the formal justice system
- % of disputes regarding women's land and property rights before the courts and alternative dispute resolution mechanisms.

Ahikire, J. 2010. *Half full or half empty glass. Decentralisation and women's land rights in Uganda*. Paper presented at the Women's Land Rights Conference, Speke Resort Munyonyo, Uganda, 4–6 October 2010.

Carfield, M. 2011. Land justice in Uganda: Preserving peace, promoting integration. *In* E. Harper, ed. *Working with customary justice systems: post-conflict and fragile states, pp. 127--143.* Rome, International Development Law Organization.

Daley, E. 2011. *Gendered impacts of commercial pressures on land.* Rome, ILC. www.landcoalition.org/sites/default/files/publication/902/mokoro_gender_web_11.03.11.pdf

Daley, E. Dore-Weeks, R. & Umuhoza, C. 2010. Ahead of the game: Land tenure reform in Rwanda and the process of securing women's land rights. *Journal of Eastern African Studies*, 4(1): 131–152.

Djusaeva, S. 2012. *Promoting rural women's rights to land: UN Women programme experience in Kyrgyzstan and Tajikistan.* Paper prepared for the Expert Group Meeting on Good Practices in Realizing Women's Rights to Productive Resources, with a Focus on Land, organized by UNWomen in collaboration with the United Nations Office of the High Commissioner for Human Rights, Geneva, 25–27 June 2012.

Englert, B. & Daley, E. 2008. Introduction. *In E. Daley and B. Englert. Women's land rights and privatization in Eastern Africa.* Woodbridge, UK, Boydell and Brewer.

FAO. 2001. *Good practice guidelines for agricultural leasing arrangements. Land Tenure Studies No. 2.* Rome.

FAO. 2002c. *Law-making in an African context: The 1997 Mozambican Land Law*, C. Tanner. FAO Legal Papers Online No. 26, Rome.www.fao.org/fileadmin/user_upload/legal/docs/lpo26.pdf

FAO. 2004. *Leasing agricultural land. Land Tenure Notes No. 1.* Rome. ftp://ftp.fao.org/docrep/fao/007/y5513e/y5513e00.pdf

FAO. 2006a. *Gender and law: Women's rights in agriculture, L. Cotula.* FAO Legislative Study No. 76, Rev.1.Rome. ftp://ftp.fao.org/docrep/fao/005/y4311e/y4311e00.pdf

FAO. 2006b. *Land tenure alternative conflict management. Land Tenure Manuals No. 2.* Rome. ftp://ftp.fao.org/docrep/fao/009/a0557e/a0557e00.pdf

FAO. 2008a. *Compulsory acquisition of land and compensation. Land Tenure Studies No. 10.* Rome. ftp://ftp.fao.org/docrep/fao/011/i0506e/i0506e00.pdf

FAO. 2009c. "We all inherited this earth", inheritance rights of widows and children in the context of AIDS: legal overview. Rome. (unpublished report)

FAO. 2011a. *Governing land for women and men – gender and voluntary guidelines on responsible governance of tenure of land and other natural resources*, E. Daley and C.M. Park. *Land Tenure Working Paper No. 19.* Rome. http://www.fao.org/docrep/014/ma811e/ma811e00.pdf

FAO. 2012b. *Voluntary Guidelines on the Responsible Governance of Tenure of Land, Fisheries and Forests in the Context of National Food Security.* Rome. http://www.fao.org/docrep/016/i2801e/i2801e.pdf

Government of Samoa. 2008. *Legislative drafting handbook.* Legislative drafting requirements approved by the Attorney General for use in Samoa from July 2008. Apia, Legislative Division of the Office of the Attorney General.

GROOTS Kenya. 2011. Complementing the state: the contribution of the "watchdog groups" in protecting women's land rights in Gatundu District, Kenya. Rome, ILC.

Guyer, J. 1987. Women and the state in Africa: Marriage law, inheritance and resettlement. *Working Papers in African Studies No. 129.* Boston, Massachusetts, USA, African Studies Centre, Boston University.

Harrington, A. & Choprah, T. 2010. *Arguing traditions. Denying Kenya's women access to land rights.* Justice for the Poor Research Report No. 2/2010. Washington, DC, World Bank.

ILC. 2010c. Women's Access to Land – Project Brief 3: Promoting women's property rights and ownership in Pakistan. Rome. (unpublished)

ILC. 2010d. Women's Access to Land – Project Brief 4: Sustainable empowerment of tribal communities through land rights movement with women partnership for securing dignity and equity. Rome. (unpublished) International Justice Mission. Our Work: What We Do.http://www.ijm.org/ourwork/whatwedo

Kapur, A. 2011. Two faces of change: The need for a bi-directional approach to improve women's land rights in plural legal systems. *In* E. Harper, ed. *Working with customary justice systems: Post-conflict and fragile states*, pp. 73–92. Rome, International Development Law Organization.

Knight, R. 2011. The community land titling initiative: An investigation into the protection of customary land claims. *In* E. Harper, ed. *Working with customary justice systems: post-conflict and fragile states*, pp. 145–170. Rome, International Development Law Organization.

Mackenzie, F. 1989. Land and territory: the interface between two systems of land tenure, Murang'a District, Kenya. Africa, 59(1): 91–109.

Mackenzie, F. 1990. Gender and land rights in Murang'a District, Kenya. *Journal of Peasant Studies*, 17(4): 609–643.

Mackenzie, F. 1993. A piece of land never shrinks: Reconceptualizing land tenure in a smallholding district, Kenya. *In* T. Bassett and D. Crummey, eds. *Land in African agrarian systems.* Madison, Wisconsin, USA, University of Wisconsin Press.

Magezi, J. 2010. *The role of case law in upholding women's land Rights.* Paper presented at the Women's Land Rights Conference, Speke Resort Munyonyo, Uganda, 4–6 October 2010.

McAuslan, P. 2010. Personal reflections on drafting laws to improve women's access to land: Is there a magic wand? *Journal of Eastern African Studies*, 4(1): 114–130.

Nakayi, R. 2010. *Is more less? Enhancing women's access to justice in Uganda.* Paper presented at the Women's Land Rights Conference, Speke Resort Munyonyo, Uganda, 4–6 October 2010.

Namutebi, M. 2010. *The Office of the Administrator General and women's land rights.* Paper presented at the Women's Land Rights Conference, Speke Resort Munyonyo, Uganda, 4–6 October 2010.

RWN. 2011. *Experiences of women in asserting their land rights: the case of Bugesera District, Rwanda.* Rome, Rwanda Women's Network, ILC.www.landcoalition.org/sites/default/files/publication/956/wlr_5_rwn.pdf

Ssonko, M. N. 2010. *The application of succession law in the realization of women's land rights in Uganda.* Paper presented at the Women's Land Rights Conference, Speke Resort Munyonyo, Uganda, 4–6 October 2010.

ULA. 2010. *The Uganda Land Alliance quarterly newsletter, Vol. 4.* Kampala.

Undeland, A. 2012. *Lessons learned and good practices in realizing women's rights to productive resources with a focus on land: Kyrgyzstan and Tajikistan.* Paper prepared for the Expert Group Meeting on Good Practices in Realizing Women's Rights to Productive Resources, with a Focus on Land, organized by UN Women in collaboration with the United Nations Office of the High Commissioner for Human Rights, Geneva, 25–27 June 2012.

UN Women. 2011. *Progress of the world's women 2011–2012: In pursuit of justice.* New York. http://progress.unwomen.org/pdfs/en-report-progress.pdf WLLA. 2010a. *The impact of national land policy and land reform in Uganda.* Accra, Women's Land Link Africa.

WOLREC. 2011. *Women's access to land and household bargaining power: a comparative action research project in patrilineal and matrilineal societies in Malawi.* Research Report No. 9. Rome, ILC, Women's Legal Resource Centre. www.landcoalition.org/sites/default/files/publication/959/wlr_9_malawi.pdf

World Bank/FAO/IFAD. 2009a. Module 4: Gender issues in land policy and administration. *In Gender in agriculture sourcebook.* Washington, DC and Rome.

YWAP. 2011. *Assisting Kayole widows in gaining control to family land, Kenya.* Research Report No. 10. Rome, ILC, Young Widows Advancement Program. www.landcoalition.org/sites/default/files/publication/951/wlr_10_ywap.pdf

MODULE

3

Institutions

LIST *of* ILLUSTRATIONS

Boxes

Checklists

Tables and figures

MODULE 3: **Institutions**

Building gender-equitable land institutions and developing capacity for responsible governance of land tenure

Introduction[6]

Governing land for women and men, fairly and responsibly and in ways that provide positive gains for society as a whole, requires the equitable inclusion of both genders in all land institutions and governance processes. There is therefore need to develop the capacity of both women and men to participate effectively in land tenure governance institutions, and the capacity of land institutions to address gender issues in the processes for governing land tenure.

Gender-equitable land tenure governance institutions are supported by the Guidelines' core principle of implementation on consultation and participation, as well as several other paragraphs in the text (Box 3.1).

BOX 3.1:
Institutions and the Guidelines

3.B.6 Consultation and participation: engaging with and seeking the support of those who, having legitimate tenure rights, could be affected by decisions, prior to decisions being taken, and responding to their contributions; taking into consideration existing power imbalances between different parties and ensuring active, free, effective, meaningful and informed participation of individuals and groups in associated decision-making processes.

5.3 States should ensure that policy, legal and organizational frameworks for tenure governance recognize and respect, in accordance with national laws, legitimate tenure rights including legitimate customary tenure rights that are not currently protected by law.... States should provide frameworks that are non-discriminatory and promote social equity and gender equality.

5.6 States should place responsibilities at levels of government that can most effectively deliver services to the people. ... States should ensure coordination between implementing agencies, as well as with local governments, and indigenous peoples and other communities with customary tenure systems.

6.1 To the extent that resources permit, States should ensure that implementing agencies and judicial authorities have the human, physical, financial and other forms of capacity to implement policies and laws in a timely, effective and gender-sensitive manner. Staff at all organizational levels should receive continuous training, and be recruited with due regard to ensuring gender and social equality.

Source: FAO, 2012b.

Land tenure governance institutions

State land institutions with responsibilities for land tenure governance include national land agencies and district and local-level land administration and management institutions. Institutions at the national, district and local levels have different roles in the processes of land tenure governance. National land agencies and ministries usually oversee policy and strategy development and implementation and provide high-level technical support. Institutions at the district and local levels concentrate more on delivering land-related services to the public, and are therefore often more directly involved in day-to-day land use and management.

These institutions interact with other types of land-related institution, such as pastoral range management institutions, watershed and irrigation management committees, customary land administration institutions, and institutions responsible for regulation of economic activities such as commercial resource exploitation (e.g. agro-industry, fishery, mining, tourism etc). Figure 3.1 illustrates some of the linkages.

Gender-equitable representation and participation in the institutions of land tenure governance includes having women as leaders in decision-making bodies and as active participants in planning processes, and may increase the likelihood of achieving gender-equitable outcomes from the processes of land tenure governance (World Bank/FAO/IFAD, 2009a: 131).

FIGURE 3.1
Interactions among institutions involved in land tenure governance

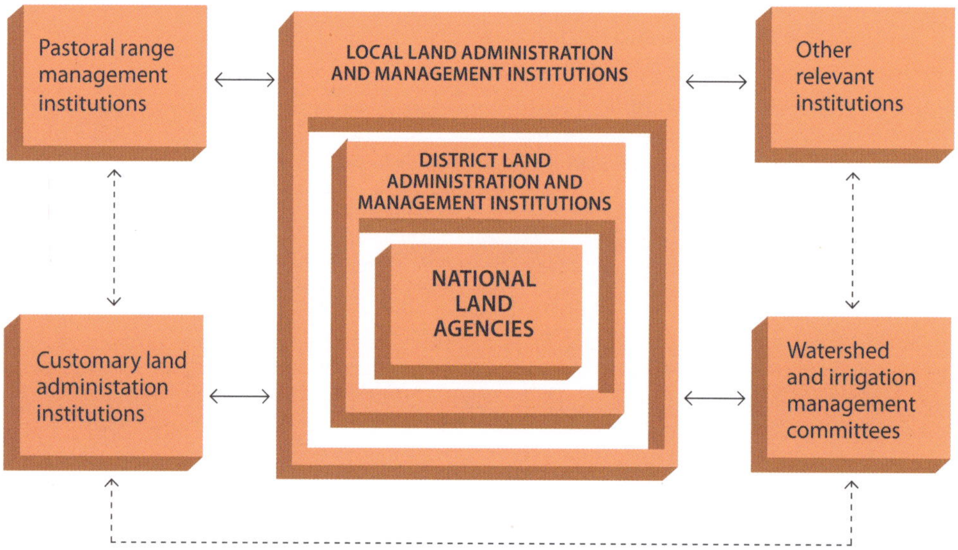

Institutional structures, linkages and operational procedures differ among countries and contexts. The measures needed to support women's representation and participation in the institutions involved in land tenure governance will therefore also vary.

In the Plurinational State of Bolivia, for example, the Andean Centre for Communication and Development, a civil society group, promotes women's participation in indigenous land management institutions through training and leadership workshops (Patricia Costas's presentation at the May 2011 technical workshop). In Rwanda, in 2008/2009, the government's National Land Centre provided training to district land officials on the duties and functions of new district and local land committees, including by informing them about the legal requirements for women's representation and participation (Elizabeth Daley, personal communication).

Other issues to consider relate to the incentives for staff and members of land institutions to promote gender equity within their institutions, and how to create willingness and commitment to support gender-equitable land institutions in general.

National land administration and management institutions

Gender-equitable representation and participation in national land administration and management institutions is still a challenge in many countries, because of a lack of opportunities, qualifications and staff capacity development to improve gender balances.

Efforts to encourage women to work in land institutions have boosted gender-equitable participation in the National Land Agency in Jamaica, as described in Box 3.2. The situation in Jamaica is helped by generally high rates of female education and literacy throughout the Caribbean; in regions where women's education and literacy rates are substantially lower than men's, the Jamaican strategy would not be sufficient on its own.

To improve the gender equity within national land administration and management institutions, there is usually need for affirmative action in employment policies, and identification of the impediments facing women, especially at senior levels. Adequate mechanisms are also needed to ensure women's representation in decision-making, planning and implementation processes affecting their land rights.

BOX 3.2:
Gender equity in national land administration

Source: Elizabeth Stair's presentation on "Land and gender: the Jamaican context" at the May 2011 technical workshop.

Jamaica's National Land Agency (NLA) is mandated to improve land administration and promote the regularization of land tenure for both women and men, although it has no specific mandate to promote gender equity in land tenure or its governance. However, the Government's Land Administration and Management Programme (LAMP) facilitates land registration at low cost for poor households, and approximately two-thirds of Jamaican households living below the poverty line are headed by women. In addition, in April 2011, NLA established gender focal points to improve the sex-disaggregation of the land data it collects.

NLA's staff is gender-balanced, with 51 percent (297) men and 49 percent (290) women; the management team is 57 percent (16) men and 43 percent (12) women. However, there are distinct gendered variations in some technical functions – only 6 percent of NLA's lawyers are men, and only 20 percent of its surveyors are women. Among land valuation staff, where men often dominate, 52 percent are women who have been encouraged to join the profession by other women land valuers. This demonstrates the value of women leaders in improving gender equity in national land administration and management institutions.

District and local-level land administration and management institutions

In many countries, land administration by local and district government is central to the implementation of land policies and programmes, with professional staff in district-level institutions acting as intermediaries between national land agencies or ministries and the grassroots level of local land committees.

Decentralization creates both opportunities and challenges for gender-equitable governance of land tenure, particularly regarding women's land rights. On the one hand, it increases the possibility for gender-equitable participation in land institutions, because it demands the participation of local communities and a wide variety of interest groups in consultative land tenure governance processes (ILC, 2009: 3).

On the other hand, decentralized land institutions may be less receptive to gender-equitable participation and representation. If decentralization is adopted, particular attention must be paid to which tasks are decentralized and mandates, roles and responsibilities must be clearly established. Also, decentralized land institutions may fail to enforce mandatory quotas designed to ensure gender-equitable participation, or facilitate women's participation at only the minimum level acceptable; cultural norms and attitudes about gender and land may be stronger at levels of governance nearer to the grassroots (Whitehead and Tsikata, 2003). Measures to promote and support women's effective participation include establishing mandatory minimum levels of women members and leaders that ensure real representation of women on district and local-level land institutions.

Box 3.4 provides examples of measures for promoting gender-equitable participation in local-level land institutions, while *Module 5* discusses gender sensitization and awareness-raising actions that support long-term change of deeply entrenched values and attitudes within institutions and beyond.

At the local level, women may also be discriminated against and excluded from processes for recording customary land rights. Ensuring greater participation of women in local-level committees that administer and manage land can improve women's chances of benefiting from the recording of customary land rights, as the case from Amhara State in Ethiopia described in Box 3.3 illustrates.

Members of local land committees are usually volunteers chosen through election or by appointment; their status in respect of statutory land institutions varies (Hilhorst, 2010). Actions for encouraging women and marginalized

BOX 3.3
Women's participation in local land governance

Sources: Hilhorst, 2010; Teklu, 2005

In *Ethiopia*'s Amhara State, field research by the International Institute for Environment and Development found that although gender equality is enshrined in the regional land policy, there are gaps in its implementation. The research involved focus group discussions with women's groups and land-use administration committees (LUACs) at the Goat level – the smallest administrative unit of local government – to gather views and experiences of the local land registration process. Separate focus group discussions and interviews were held with female-headed households and desk officers of the Natural Resource and Land Use Administration Bureau at the Woreda (district) level.

The research found very few women on village and Goat-level LUACs, despite instructions from the Woreda level that village LUACs have at least two women members. Focus groups revealed that women did not participate in the meetings to elect LUAC members because they did not attend meetings, lacked information on the timing and location of meetings, or saw the election of LUAC members as a task for men. There was also inadequate staffing of Woreda-level land-use administration desks (LUADs) and no gender focal point to oversee the land registration process. The LUAC members had different understandings and interpretations of the law, partly because of their limited education, but also because Woreda desk officers lacked training and follow-up capacity.

Most of the women consulted in the research believed strongly that involvement in land registration was a task for men, and that women's participation in local committees for the registration process would make no difference. However, this excerpt about a woman on a Goat LUAC highlights the importance of women committee members:

> *She is 34 years old, married, and educated to grade seven. The work was challenging for her in the beginning given the general perception of women not being able to handle such tasks. However, she has proved herself competent and even better than some men on the committee, as witnessed by other male committee members and the Woreda LUAD officer. In her experience, women tend to feel freer telling their problems to other women rather than to a man. She also remarked that women tend to be more neutral and abide by the law in implementing such sensitive policies. She further lamented that a considerable number of men had tried to exclude their wives' names from the registration. In several cases, some of the LUAC members colluded with these husbands, whereas she believed that any women on the committee would stand up for women's rights.*

groups to participate in local institutions responsible for land tenure governance include convening women separately, to help them develop confidence on the issues in preparation for participating in full committee meetings and wider debates (Rachael Knight's presentation at the May 2011 technical workshop).

However, care is needed when mandating women's involvement in district and local-level institutions. It is not always effective to create new leadership structures, or to mandate minimum levels of women's representation and participation in land institutions. Awareness of gender equity issues is just as important, as the project in Liberia described in Boxes 2.7 and 2.14 illustrates. Although women's involvement was not mandated in Liberia, project outcomes included the election of women as town chiefs through their own improved awareness of gender equity issues. When women's representation on local land institutions is obligatory, it may be best to leave the women and men in communities to decide themselves how they will comply (Knight, 2011).

CHECKLIST 3.1: **Ensuring gender-equitable land institutions**

- ✓ Create institutional mechanisms – including those based on membership – that enable women both to support and to have a voice in local government bodies.
- ✓ Encourage the use of quotas of women and men members in national land agencies and district and local land committees.
- ✓ Establish quotas of women and men for a quorum at land institution meetings.
- ✓ Sensitize all members of land institutions on the importance of gender equity in their processes and composition.
- ✓ Encourage women's participation in meetings about land issues and in the land institutions that represent them, for example through training and leadership workshops to boost their capacity.

REMEMBER!
Mandatory representation of women in the institutions of land tenure governance is an important start, but it is insufficient for ensuring women's active and effective participation alongside their male colleagues. Sensitization and capacity building for all members of land institutions is needed to support gender-equitable governance of land tenure.

Land-related institutions

Institutions such as local watershed and irrigation management committees, pastoral range management institutions and customary land administration institutions often deal with issues at the interface of land and other natural resources, such as access to water for domestic use, pastoral water rights, customary access rights to water for irrigation, and access to rangeland for pastoral grazing (FAO, 2011a: 25). Because of these close links, policy reforms that have attempted to regulate different natural resources separately have often had unforeseen consequences on local land tenure and use patterns, including negative impacts for women (Adams, Berkoff and Daley, 2006). Ensuring that the local institutions dealing with watershed, irrigation and pastoral range management are gender-equitable can help to avoid some of these unforeseen consequences. Box 3.4 gives examples of lessons learned from efforts to encourage gender-equitable participation in land-related institutions.

BOX 3.4
Gender-equitable participation in land-related institutions

Sources: Seeley, Batra and Sarin, 2000; FAO, 2006a: 45–46; World Bank/FAO/IFAD, 2009c: 451–452 and 456–457; Flintan and Cullis, 2010; Agarwal, 2009.

In *India*, the Watershed Guidelines of 1994 encourage greater participation of women and marginal groups in watershed development. The guidelines call for watershed projects to establish user groups and self-help groups that include women or are for women only. However, gender-equitable participation has been difficult to achieve, and rather than being recognized as full members of watershed management groups, women are generally viewed as being there to fill quotas. This is in part because watershed management in India focuses on the agricultural land in watersheds, which is controlled by men. In 1999, in response to criticism that its watershed development programme was for men only, the state government of Andhra Pradesh formally linked 1200 women's self-help groups in Mahaboobnagar to the programme, and representatives from these groups, drawn from different social and income groups, now represent women on the watershed committee.

In *Ethiopia*, participatory approaches to dryland management in pastoral areas have opened new spaces for more gender-equitable participation. Much of this success is due to resource mapping through a gender-sensitive community process in which mapping and information-gathering exercises are carried out with women and men separately. This ensures that the views and perspectives of all resource users are included.

In *India* and *Nepal*, evidence suggests that overcoming constraints to women's participation in forest governance is likely to have a beneficial impact on conservation outcomes. For instance, one study found that community forest management groups with high proportions of women on their executive committees brought significantly greater improvements in forest condition. Groups with all-women executive committees in Nepal also had better forest regeneration outcomes than other groups, despite managing much smaller and more degraded forests.

In *Egypt*, the World Bank's Matruh Resources Management Project seeks to break the cycle of natural resource degradation and poverty in Bedouin communities. The project works with community groups to define the needs of both women and men and ensure that both participate in preparing and implementing local resource management plans. To enable the community groups to address gender issues effectively, project staff received gender training, and women extension agents in each sub-project area work directly with women.

The Farmer-Centred Agricultural Resource Management (FARM) programme – implemented by FAO and supported by the United Nations Development Programme (UNDP) in *China*, *India*, *Indonesia*, *Nepal*, the *Philippines*, *Sri Lanka*, *Thailand* and *Viet Nam* between 1993 and 1998 – promoted household food security and the sustainable use and management of natural resources in agriculture in ecologically fragile, rainfed areas. The programme encouraged women's participation at all levels of decision-making processes and other activities. FARM adopted a participatory assessment planning approach that incorporated a practical gender analysis tool for examining the activities, problems, knowledge and access to natural resources of both women and men, and carried out training of trainers under FAO's Socio-economic and Gender Analysis Programme (SEAGA). The outcome was greater awareness and understanding of social equity issues among community members, with women beginning to play important roles in decision-making and leadership.

In *Pakistan*, the local NGO, Pattan, supports underrepresented community organizations and develops gender-equitable local institutions, to increase communities' capacity to cope with natural disasters such as floods. Pattan worked with communities to organize representative, democratic fora called Pattan Dehi Tanzeems (PDTs) for collective decision-making. Barred by local tradition from joining the PDTs, women formed their own. When the 1992 floods destroyed many villages, Pattan initiated a project to rehabilitate houses, with women participating in the project's PDTs. In Pakistan, women maintain traditional (kacha) housing, but the project involved them in the design and construction of improved (pakka) housing. Pattan introduced the concept of joint ownership of the new housing, and provided households with loans, which women took responsibility for repaying. Initially, men objected to women's assumption of this responsibility, but eventually they saw the value of joint ownership. Experience of the housing project has given women confidence to take collective action in other projects.

Working with customary land institutions

In many countries, customary institutions govern all aspects of people's social, cultural, economic and political lives and play a central role in governing land, water and grazing rights. Customary land administration institutions should therefore be supported as part of a broad and pragmatic approach to change (Daley and Englert, 2010).

Male-dominated, discriminatory and exclusive customary institutions should be assisted in overcoming these obstacles and accepting principles of gender equity. For instance, traditional leaders and women at the grassroots can be involved in distinguishing between cultural practices that support the land rights of women and those that do not. Other options include providing community training for men and women leaders, show-casing good practices, and promoting peer exchanges (WLLA, 2010a).

As noted in *Module 2*, women sometimes find it easier to approach local customary leaders than formal institutions, especially when the formal institutions are new (FAO, 2011a: 17). Women may have their own separate customary institutions, which can be incorporated into responsible gender-equitable land tenure governance without the need to create new institutions (Flintan, 2010b).

Customary land administration institutions can have an important role in influencing changes towards more gender-equitable tenure practices, as the cases from Uganda and Malawi described in Box 3.5 illustrate.

BOX 3.5:
Working with costumary institutions

Sources: WOLREC, 2011; ULA, 2010; Adoko, 2011; Akin, no date.

The Land Equity Movement of *Uganda* (LEMU) was established in 2003 to advance gender-equitable customary land tenure governance in northern and eastern Uganda. It involves local customary authorities in creating consensus and encouraging the harmonization of customary and statutory systems, to promote security of land rights. LEMU's awareness-raising activities include:

- documenting customary law in codes of practice for distribution to the formal courts;
- supporting communities in demarcating the boundaries of their landownership and use rights through local methods such as tree planting and sketch maps;
- training police on customary land rights and supporting the enforcement of judgments;
- training university students in customary land tenure and creating opportunities for them to mediate in real-life land disputes.

At the request of district governments and the Ministry of Lands, LEMU trains customary leaders and local land committee members on statutory and customary land laws and conflict mitigation, and on setting up area land committees for local land administration.

In *Malawi*, an action research study by the Women's Legal Resource Centre (WOLREC), funded by the International Land Coalition (ILC), worked with chiefs in bringing social change to allow widows' access to land. The chiefs recognized that the payment of bride-wealth in patrilineal areas provides space for widows to negotiate access to land. Widows are usually dispossessed, mainly because land scarcity causes the families of their deceased husbands to challenge the customary notion that women acquire joint ownership of their husbands' land on marriage. Following the study's efforts, some chiefs started to discredit land dispossession of widows at public meetings, funerals and other village gatherings. First, however, the study had to overcome limitations that included chiefs' initial reluctance, fear of higher authority, and lack of knowledge about statutory land laws.

CHECKLIST 3.2:
Working with customary and local institutions

Sources: Judy Adoko's paper on "Constructive ways of working with customary institutions in support of women's land rights" at the May 2011 technical workshop; Flintan, 2010a; 2010b.

✓	Research customary tenure practices and institutions, and raise awareness about them among national policy- and law-makers. Seek understanding of how decisions about land tenure governance are made and translated into actions and results.
✓	Identify customary practices, processes and institutions that support gender-equitable land tenure governance, and seek to eliminate those that are discriminatory or exclusionary.
✓	Sensitize customary leaders and members of local land-related institutions on the importance of gender-equitable land tenure governance, and support them in adapting to new governance processes. Do not assume that all customary leaders are against change.
✓	Support land administration by customary institutions, and ask governments to hold them accountable for gender equity.
✓	Incorporate women's existing customary institutions and fora into local land tenure governance arrangements.

REMEMBER!
Existing customary institutions are often the best place to start in pursuing responsible, gender-equitable governance of land tenure at the local level.

Other institutions

The judiciary, with its courts and offices at the central and local levels across the country, also has a role in supporting gender-equitable land tenure governance. Judges, prosecutors, police chiefs and other personnel must be aware of the law's stipulations on gender equality in land and property rights and of the constraints that different groups of people face in trying to protect their rights. As noted in *Module 2*, supporting women's participation in the judiciary helps to improve women's access to justice.

Producer and farmers' organizations can support gender-equitable land tenure governance at the grassroots level. These institutions often serve as intermediaries between local communities, and national and global policy-making bodies and government ministries, and engage in useful land-related advocacy and information dissemination activities at various levels. Supporting women's representation in rural organizations facilitates gender equity in the management and decision-making processes of these institutions, and thus in wider land tenure governance processes (FAO, 2011a: 11).

Gender equality and participation of every person - women and men - in decision-making processes and institutions is a human right, and evidence suggests that enabling local women's participation alongside men in all land-related institutions can also improve the overall gender equity of outcomes for land tenure governance. Box 3.4 provides examples of good practice in this regard.

Capacity development for gender-equitable land tenure governance institutions

Specific capacity development measures are needed to address women's lack of access to higher education and lack of incentives for undertaking scientific and technical studies, which hamper women's entry into some of the technical professions in the land sector. More general capacity development is also needed across the institutions of land tenure governance, to increase knowledge and understanding of gender issues. Box 3.6 provides examples of good practice in gender-equitable institutional capacity development.

BOX 3.6:
Capacity building for gender-equitable land tenure governance

Sources: World Bank/FAO/IFAD, 2009b: 457 and 463–465; ADB, 2006; Khudayberdiyeva, 2009; Djusaeva, 2012.

In *India*, between 2001 and 2009, the World Bank-funded Karnataka Watershed Development Project aimed to improve the productive potential of watersheds in seven districts of Karnataka State through local participation in community natural resource management, including by enhancing women's opportunities to participate in decision-making. The project developed partnerships with government technical specialists, NGOs, communities, local authorities and research organizations, and formed community-based organizations, such as groups of women and the landless, which were formally linked to the watershed programme. The creation of women-only fora supported by men encouraged shared decision-making and even asset ownership. Benefits were reflected in better health, education and overall standards of living for participating families.

In *Tajikistan*, a UN Women-supported programme promoted the adoption of a mechanism for providing the Land Agency with citizens' feedback on ongoing land reform. The mechanism used two-way communication between the district Land Agency and rural citizens, with emphasis on efforts to remove cultural and social stereotypes, which are still strong in rural areas. Piloting of the feedback mechanism, monitored through citizens' report cards, showed a strong need for further strengthening of local governments' capacities for rights-based and gender-sensitive service provision. As a result, more than 650 local government officials and staff from the Land Agency, the Women's Committee, NGOs and informal rural community groups attended training workshops on gender aspects of the land reform process. This improved trainees' involvement in the land restructuring process, from better skills in dealing with farmers' concerns, to greater participation in awareness campaigns.

CHECKLIST 3.3:
Building women's capacity

Sources: Discussions of the working group on institutions at the May 2011 technical workshop; World Bank/FAO/IFAD, 2009b:458.

- ✓ Train women in public speaking and debating techniques to boost their confidence.
- ✓ Train women on relevant policies and laws so they can contribute effectively in meetings of the institutions they belong to.
- ✓ Train women and men on institutional procedures and processes, particularly on provisions for supporting and improving gender equity in land tenure governance.
- ✓ Ensure that institutional operating procedures consider women members' needs, time commitments, mobility restrictions and cultural norms, to maximize the impact of women's participation. This might include making provision for child care during institution meetings and activities.

REMEMBER!
Women members of land institutions may need special training and support to carry out their roles effectively, particularly where they have limited experience of participating in such institutions.

CHECKLIST 3.4: **Developing institutional capacity for gender equity**

Source: Seeley, Batra and Sarin, 2000.

✓	Recruit women to top positions in land institutions – to set examples of institutional leadership on gender equity – and provide appropriate support for them.
✓	Sensitize and train staff and members of all the institutions involved in land tenure governance – including the police, judges and customary authorities – on gender issues.
✓	Encourage private sector organizations in the land sector to address gender equity in their work with land institutions at all levels. Do not rely on governments and large multinationals to take the lead in changing everyday institutional practice at the local level.
✓	Train the women field staff of land-related projects and women land officials in facilitating women's participation in local land tenure governance institutions.
✓	Pursue participatory processes and innovations at local levels of planning and decision-making, to build capacity from the grassroots up.
✓	Ensure that women and men members of land institutions are given equally important roles and responsibilities.
✓	Recognize that women from different socio-economic backgrounds, religions, castes, etc. may have different interests in participating and decision-making in land institutions. Institutional structures and processes must take this into account to ensure equitable governance of land tenure for all women and men.

REMEMBER!
Institutions evolve; building institutional capacity to address gender equity in land tenure governance must therefore be an evolutionary process.

Gender-equitable representation in technical professions

Land administration has traditionally been a male-dominated sector – particularly in more technical professions such as surveying. This has led to perceptions of gender equity as requiring only that women are employed somewhere in land administration institutions and projects, regardless of the roles they are assigned (World Bank, 2005: 18).

Young women seeking to enter the land sector as technical professionals often have to overcome:

- physical and personal constraints to spending prolonged periods in the field in often basic living conditions, as required in professions such as surveying;
- lower levels of technical education than young men, as families give preference to males in pursuing higher education, particularly in scientific subjects;
- indirect job discrimination because of the stereotyping of girls and young women in domestic, reproductive and subservient roles, and because of gender-based discriminatory practices in recruitment processes;
- less access to information channels and job search mechanisms than young men, including through professional networks and associations.

Women generally tend to be better represented in office work for land administration. For example, women have an acknowledged important place as customer service agents in land registration offices. It is often easier for a woman client to approach another woman at a front desk, so it is important to ensure a gender balance in staffing the office.

In many countries women are also involved in articulating how land information systems work, including through the development and deployment of tools in the private sector. They are often also well represented as users of land information systems and databases, for example in many clerical and data entry roles.

However, concerted actions are needed to improve gender-equitable participation in the technical aspects of land administration, particularly in ensuring a better gender balance between office and fieldwork. For example, measures are needed to raise awareness and provide information on jobs, making it clear that they are open to both genders, and efforts to attract more young women into technical professions and field-based work should explain the pros and cons and facilitate practical field experience. The case study from Ghana described in Box 3.7 provides a good practice example in supporting scientific and technical education for women and girls. Box 3.8 illustrates the situation at the opposite end of the spectrum in the Arab region.

BOX 3.7:
Supporting scientific and technical education for girls

Sources: Rebecca Sittie's presentation on "Gender issues in managing and implementing land registration in Ghana" at the May 2011 technical workshop; Rebecca Sittie, personal communication.

With support from the Ministry of Women and Children's Affairs, *Ghana*'s Ministry of Education runs "clinics" during the long school vacations in all ten regions. Women involved in science and maths attend these clinics to speak to girls aged 13 to 16 years, encouraging them to pursue science and technical subjects at school and beyond. Women role models also give motivational talks at primary schools, to encourage girls to aim high. In parallel, the women's wing of the Ghana Institution of Surveyors organizes visits to girls in high school, to introduce the surveying courses available at the university level.

As a result of these efforts, and of the government's general policy of promoting girls' education, in 2005, 40 out of 110 new graduates from the Department of Land Economy at Kwame Nkrumah University of Science and Technology were women. Subsequent years have recorded similar proportions of female enrolments in land economy courses.

BOX 3.8:
Challenges to gender equity in land administration activities and technologies

Sources: Dalal Alnaggar's presentation on "Developing gender-equitable land survey and technologies in the Arab region" at the May 2011 technical workshop; Dalal Alnaggar personal communication.

In the 22 countries of the Arab world gender discrimination has complicated religious, cultural, social, political and economic dimensions. Men involved in technical aspects of the land sector in these countries often feel that gender discrimination protects women because, for example, fieldwork is considered inappropriate for women, and the equipment too heavy. At the same time, some women – especially in rural areas – are afraid of moving beyond their traditional and often secluded gender roles because of the potential impact on their reputations. Women lack time because of their heavy domestic burdens, and their mobility is also often limited – they may not be permitted to travel or go out alone. Education is heavily gender-segregated, and women who work often choose local jobs that do not conflict with their domestic burdens and social reputations. Across the Arab world, only 18 percent of married women are part of the paid labour force, and women hold only 2 percent of high-level positions in science sectors.

In *Egypt*, *Jordan*, *Lebanon*, *Morocco* and *Tunisia* women can work in the land sector's technical community, but not all jobs are open to both men and women; some jobs requiring advanced qualifications are reserved for men. Across the Arab world there are big gender gaps at the decision-making levels in scientific and technical areas of both government and the private sector. In Egypt, for example, in the decade 2000–2010, women comprised about one-third of the students on university science courses, but in 2011 only 50 women had masters' degrees in surveying and only five had Ph. D.s ; there were no postgraduate women surveyors in any other Arab country.

CHECKLIST 3.5:
Actions for countries with severe gender disparities

✓	In contexts where severe gender disparities exist, identify the main challenges to gender equity in the technical aspects of land administration.
✓	Create a databank of scientific and technical jobs, and encourage women to apply.
✓	Support women who are specializing in information technology, valuation and standardization.
✓	Support scientific and technical education and capacity building measures for girls and young women.
✓	Establish minimum quotas for women's participation in technical roles in land administration agencies and departments.
✓	Support women surveyors and other technical professionals in working with international agencies, governments and the private sector, by using appropriate technology and information, including new technologies that are more gender-neutral.
✓	Raise the standard and coverage of scientific and technical education in rural areas, with special attention to the education of girls.
✓	Raise awareness on the importance of women identifying their own rights and responsibilities. Focus on abolishing the belief that work is demeaning for women and conflicts with their domestic burdens and social reputations.

REMEMBER!
Targeted measures will be needed to support women's participation in land administration in countries where severe gender disparities are deeply entrenched in cultural and religious beliefs.

To improve the gender balance among land sector professionals and support the equitable participation of both women and men in land administration there is a worldwide need to develop technologies and systems that can be used by a wider group of people, and to revise the qualifications needed for certain technical jobs, to reduce access barriers, especially for women. Surveying and measurement techniques should also be simplified wherever possible so they can be managed in the field by lower-qualified para-surveyors (FAO, 2011a: 10). The increasing use of participatory and low-cost methods of land surveying and registration is already having a significant positive impact on gender equity in land tenure governance.

CHECKLIST 3.6:
Improving gender equity in land administration

✓	Support effective and sustainable access to technical areas of the land sector for both women and men, including through affirmative action in the short to medium term.
✓	Collect and disseminate gender-disaggregated data in all land administration institutions and projects, including on staffing.
✓	Organize and develop women's technical capacity in the land sector through professional education, training and professional networks, including associations of women surveyors.
✓	Identify champions and promote women mentors and role models.
✓	Develop on-the-job training to build the capacity of scientific and technical institutions to support women's career development, for example through professional practical experience in the field after university.
✓	Facilitate women's mobility among different technical roles in land administration, to improve their understanding of the full spectrum of technical land administration work as they progress up the professional ladder.
✓	Organize gender training, capacity building and awareness raising for both women and men professionals in the land sector.
✓	Develop a strategy for incorporating the concerns and experiences of both women and men into the design, implementation, monitoring and evaluation of projects in all technical aspects of land administration.
✓	Provide adequate support and facilities for women to work in the field.

REMEMBER!
Gender equality in land administration will not be achieved unless women can participate in activities, technologies and systems on a par with men.

Summary of key messages on institutions

Land institutions with gender-equitable representation and participation are better at governing land for both women and men.

There is need for raising the public's awareness of the importance of effective participation by both women and men in land tenure governance institutions at all levels, from the national to the district.

Decentralization processes create both opportunities and challenges for gender-equitable governance of land tenure.

Institutuions such as local watershed, irrigation and pastoral range management institutions also have roles in responsible, gender-equitable governance of land tenure.

Customary land administration institutions can have a very important role in influencing changes towards more gender-equitable customary practices.

Specific capacity-building measures may be needed to support women members of land institutions at all levels. Institutional capacity to address gender equity in land tenure governance should be continually reviewed and developed.

Specific capacity-building measures may be needed to support women members of land institutions at all levels. Institutional capacity to address gender equity in land tenure governance should be continually reviewed and developed.

There is need for gender-equitable participation in technical aspects of land tenure governance – land administration activities, technologies and systems – to accommodate the interests of all people – women and men.

Suggested monitoring indicators for institutions

- % of women and men in central land administration and management institutions
- % of women and men in district land administration and management institutions
- % of women and men in local land administration and management institutions
- % of women and men in customary land institutions and other land-related local institutions
- number of land officials receiving gender sensitization and training
- number of customary land institutions and other land-related local institutions trained and sensitized on gender and land issues
- number of initiatives for removing barriers to land justice for women
- % of women and men receiving legal support in land cases in the formal justice system
- % of disputes regarding women's land and property rights before the courts and alternative dispute resolution mechanisms.

Notes

[6] Debra Fletcher, Rachael Knight, Martha Osorio, Elizabeth Stair, Josephine Stowers-Fiu, Mika Törhönen and Margret Vidar provided substantial inputs to this module during the May 2011 technical workshop at FAO Headquarters, including through discussions of the working group on institutions.

Adams, M., Berkoff, J. & Daley, E. 2006. *Land-water interactions: opportunities and threats to water entitlements of the poor in Africa for productive use.* Human Development Report Occasional Paper No. 2006/20. Washington, DC, UNDP. http://hdr.undp.org/en/reports/global/hdr2006/papers/adams%20martin%20et%20al.pdf

ADB. 2006. *Republic of Uzbekistan: land improvement project.* Report and Recommendation of the President to the Board of Directors, Proposed Loans and Technical Assistance Grant. Manila, Asian Development Bank. www2.adb.org/documents/rrps/uzb/37536-01-uzb-rrp.pdf

Agarwal, B. 2009. Gender and forest conservation: The impact of women's participation in community forest governance. *Ecological Economics*, 68(11): 2785–2799.

Alden Wily, E. 2003. *Governance and land relations: A review of decentralisation of land administration and management in Africa.* London, IIED. http://pubs.iied.org/pdfs/9304IIED.pdf

Basangwa, S.K. 2010. *Women in land administration.* Paper presented at the Women's Land Rights Conference, Speke Resort Muynonyo, Uganda, 4–6 October 2010.

Daley, E., Dore-Weeks, R. & Umuhoza, C. 2010. Ahead of the game: Land tenure reform in Rwanda and the process of securing women's land rights. *Journal of Eastern African Studies*, 4(1): 131–152.

Daley, E. & Englert, B. 2010. Securing land rights for women. *Journal of Eastern African Studies*, 4(1): 91–113.

Djusaeva, S. 2012. *Promoting rural women's rights to land: UN Women programme experience in Kyrgyzstan and Tajikistan.* Paper prepared for the Expert Group Meeting on Good Practices in Realizing Women's Rights to Productive Resources, with a Focus on Land, organized by UNWomen in collaboration with the United Nations Office of the High Commissioner for Human Rights, Geneva, 25–27 June 2012.

FAO. 2006a. *Gender and law: Women's rights in agriculture,* L. Cotula. FAO Legislative Study No. 76, Rev.1.Rome. ftp://ftp.fao.org/docrep/fao/005/y4311e/y4311e00.pdf

FAO. 2011a. *Governing land for women and men – gender and voluntary guidelines on responsible governance of tenure of land and other natural resources,* E. Daley and C.M. Park. *Land Tenure Working Paper No. 19.* Rome. http://www.fao.org/docrep/014/ma811e/ma811e00.pdf

FAO. 2012b. *Voluntary Guidelines on the Responsible Governance of Tenure of Land, Fisheries and Forests in the Context of National Food Security.*Rome. http://www.fao.org/docrep/016/i2801e/i2801e.pdf

Flintan, F. 2010b. Sitting at the table: Securing benefits for pastoral women from land tenure reforms in Ethiopia. *Journal of Eastern African Studies*, 4 (1): 153–178.

Flintan, F. 2010a. *Learning by doing: Working towards participatory rangeland management (PRM) in pastoral areas through ELMT/ELSE.* Washington, DC, USAID.

Flintan, F. & Cullis, A., compilers. 2010. *Introductory guidelines to participatory rangeland management in pastoral areas.* Ethiopia, Save the Children USA, FAO and ECHO. www.cop-ppld.net/fileadmin/user_upload/cop-ppld/items/introductory%20guidelines%20prm.pdf

Hilhorst, T. 2010. Decentralization, land tenure reforms and local institutional actors: building partnerships for equitable and sustainable land governance in Africa. *Land Tenure Journal*, 1(10): 35–59.

ILC. 2009. *Advancing the monitoring of land governance for ensuring impact on poverty reduction.* Paper presented atLand Governance in Support of the MDGs: Responding to New Challenges, World Bank, Washington, DC, 9–10 March 2009.

Khudayberdiyeva, M. 2009. *Uzbekistan Land Improvement Project.*Tashkent, ADB Uzbekistan.

Knight, R. 2011. The community land titling initiative: An investigation into the protection of customary land claims. In E. Harper, ed. *Working with customary justice systems: Post-conflict and fragile states,* pp. 145–170. Rome, International Development Law Organization.

Seeley, J., Batra, M. & Sarin, M. 2000. *Women's participation in watershed development in India.* Gatekeeper Series No. 92. London, IIED. http://pubs.iied.org/pdfs/6347IIED.pdf

Teklu, A. 2005. *Research report 4 – land registration and women's land rights in Amhara Region, Ethiopia.* London, IIED.

ULA. 2010. *The Uganda Land Alliance quarterly newsletter, Vol. 4.* Kampala.

Whitehead, A. &Tsikata, D. 2003. Policy discourse on women's land rights in sub-Saharan Africa: The implications of the re-turn to the customary. *Journal of Agrarian Change*, 3(1–2): 67–112.

WLLA. 2010a. *The impact of national land policy and land reform in Uganda.* Accra, Women's Land Link Africa.

WOLREC. 2011. *Women's access to land and household bargaining power: a comparative action research project in patrilineal and matrilineal societies in Malawi.* Research Report No. 9. Rome, ILC, Women's Legal Resource Centre. www.landcoalition.org/sites/default/files/publication/959/wlr_9_malawi.pdf

World Bank. 2005. *Gender issues and best practices in land administration projects: A synthesis report.* Washington, DC. http://siteresources.worldbank.org/intard/resources/gender_land_fulltxt.pdf

World Bank/FAO/IFAD. 2009a. Module 4: Gender issues in land policy and administration. *In Gender in agriculture sourcebook. Washington, DC and Rome.*

World Bank/FAO/IFAD. 2009b. Module 10: Gender and natural resources management. *In Gender in agriculture sourcebook.* Washington, DC and Rome.

MODULE

4

Technical issues

LIST *of* ILLUSTRATIONS

Boxes

Checklists

Tables and figures

MODULE 4: **Technical issues**

Improving gender equity in land administration approaches, methods and technologies

Introduction[7]

Land administration requires consideration of two main gender issues: *i)* how adequately the design of land administration activities addresses women's needs, interests and concerns; and *ii)* how women can participate equally in land administration activities.

The need to address gender equity in land administration activities, technologies and systems is directly supported throughout the Guidelines (see examples in Box 4.1).

BOX 4.1:
Gender equity, technical issues and the Guidelines

Securing and allocating land rights
7.4 States should ensure that women and men enjoy the same rights in the newly recognized tenure rights, and that those rights are reflected in records. … Locally appropriate approaches should be used to increase transparency when records of tenure rights are initially created, including in the mapping of tenure rights.
10.3 Whenever States provide legal recognition to informal tenure, this should be done through participatory, gender-sensitive processes, having particular regard to tenants.
11.6 States should establish safeguards to protect the legitimate tenure rights of spouses, family members and others who are not shown as holders of tenure rights in recording systems, such as land registries.
7.1 When States recognize or allocate tenure rights to land, fisheries and forests, they should establish, in accordance with national laws, safeguards to avoid infringing on or extinguishing tenure rights of others, including legitimate tenure rights that are not currently protected by law. In particular, safeguards should protect women and the vulnerable who hold subsidiary tenure rights, such as gathering rights.
Land rights registration
17.3 States should strive to ensure that everyone is able to record their rights and obtain information without discrimination on any basis.
Regulated spatial planning
20.2 States should develop through consultation and participation, and publicize, gender-sensitive policies and laws on regulated spatial planning.
20.3 States should ensure that regulated spatial planning is conducted in a manner that recognizes the interconnected relationships between land, fisheries and forests and their uses, including the gendered aspects of their uses.
Land restitution
14.4 States should develop gender-sensitive policies and laws that provide for clear, transparent processes for restitution. Information on restitution procedures should be widely disseminated in applicable languages. Claimants should be provided with adequate assistance, including through legal and paralegal aid, throughout the process.

Source: FAO, 2012b.

Redistributive reforms

15.5 Where States choose to implement redistributive reforms, they should clearly define the objectives of reform programmes and indicate land exempted from such redistribution. The intended beneficiaries, such as families including those seeking home gardens, women, informal settlement residents, pastoralists, historically disadvantaged groups, marginalized groups, youth, indigenous peoples, gatherers and small-scale food producers, should be clearly defined.

General land administration and management issues[8]

Some of the common obstacles to gender-equitable governance of land tenure stem from deeply entrenched values of societies (Renée Giovarelli's presentation at the May 2011 technical workshop):

- Worldwide, most land administration staff and project implementers are men; as a result women often do not participate in, or are not even present at, project activities on the ground such as land demarcation or surveying (see *Module 3*).
- Women are often not included in the design of projects or the benefits of project activities, such as access to credit or community services.
- Women often have less knowledge than men about the land tenure and governance issues affected by a project, or about the potential implications of projects for existing land tenure arrangements and governance practices.
- Land administration staff and project implementers on donor-funded projects may not be fully aware of the social and cultural aspects of land rights in the areas where they are working.

It is important to ensure that all land administration activities mainstream gender-equity issues by considering the whole range of gender issues during project design, implementation, monitoring and evaluation (World Bank/FAO/IFAD, 2009a: 130–132). The issues to be addressed will depend on the specific task and country context, but there are many examples of good practice to draw on, some of which are described in Box 4.2.

BOX 4.2:
Good practice for gender equity in land projects

Joint titling

In *India*, homestead land purchase programmes help landless rural families to acquire small plots for secure housing and small-scale home-based enterprises controlled by women. Beneficiaries' full participation at all stages of these programmes, from the identification of suitable land to the design of land development plans, has contributed to successful implementation. Many of the programmes issue joint land titles to husbands and wives, sometimes listing the wife's name first; others issue land titles to women only. Success is facilitated by awareness of gender equity issues among local communities and authorities.

In *Cambodia*'s Supplemental Land Administration Project of 2002–2007, 78 percent of new titles were issued jointly to husbands and wives. The project aimed to improve women's overall socio-economic status. To promote long-term sustainability, it was linked to other development projects, such as those for securing women's access to credit and extension services. A social assessment conducted prior to project implementation applied a gender perspective to land rights, and gender concerns were integrated into the project design. Women's participation was also emphasized in both the design and implementation phases. The project conducted a comprehensive public outreach campaign, with frequent gender-awareness training sessions involving both women and men, including land administration and management officials at all levels. Women were specifically targeted because of their higher rates of illiteracy and general lack of education about land rights. The project also hired gender advisers to work with the implementation team.

In *Ethiopia*, following issuance of the Land Use and Administration Proclamation in Amhara Region in 2000, the regional Natural Resource and Land Use Bureau developed an implementation strategy and guidelines establishing the responsibilities and accountabilities of different actors in land rights registration. The guidelines stipulated that joint registrations of land title have to include the names and photographs of both the wife and the husband on the certificate, and that both spouses have to sign for the title certificate before it is given to the household. These stipulations help to ensure gender equity in land rights registration, although districts require further support in mainstreaming gender issues into their operations and designing programmes to improve gender equity at the grassroots level.

In *Honduras*, the Guayape Valley Agricultural Development Project, funded by the Inter-American Committee for Agricultural Development, worked with the Honduras Titling Agency to include wives' names on land titles, despite limitations in the relevant legislation and inheritance and marital property practices that excluded women. The project was successful because it reviewed titling procedures to make them more accessible to and inclusive of women, and because it held gender training for project, government titling, and public and local authority staff.

Public education and awareness raising

In *Jamaica*, the National Land Agency's (NLA's) work in public education supports gender-equitable governance of land tenure. For instance, NLA organizes road-shows around the country, usually on market days, when women from rural areas are likely to be in town. At these road-shows, NLA provides information and makes presentations to women, in groups and individually. Topics include the benefits of making a will, registering their own names on their partners' property, regularizing their existing tenure through registration, and titling any land they buy in their own right. Staff of Jamaica's Land Administration and Management Programme attend the sessions. NLA provides information via brochures, other printed material and the Internet.

Participatory mapping

In *Mozambique*, FAO's Participatory and Negotiated Territorial Development (PNTD) approach has been used in the delimitation of community land. Community members, including separate groups of women and men, produce maps during a participatory consultation process. A final version of the map is then agreed on and recorded in the cadastre after discussion with neighbouring communities. FAO has recently integrated the PNTD approach and its Socio-Economic and Gender Analysis (SEAGA) approach into guidelines on improving gender equity in territorial issues, to promote the use of gender-sensitive participatory methodologies and tools in negotiation processes concerning access to and use of land and natural resources.

Sources: Elizabeth Stair's presentation on "Land and gender: the Jamaican context" at the May 2011 technical workshop; Giovarelli and Lastarria-Cornhiel, 2006; World Bank/FAO/IFAD, 2009a:139–140 and 145; Teklu, 2005; FAO, 2012a.

CHECKLIST 4.1: **Promoting gender equity in land administration projects**

- ✓ Develop understanding of the local tenure system and the provisions it makes for women. Be aware that most of this information will not be recorded in writing.
- ✓ Check limitations in the respective laws, policies and regulations concerning gender equality.
- ✓ Conduct social assessments prior to project implementation, including family structure; women's representation in local affairs, decision-making power and roles in families and society; and inheritance, marital and divorce patterns.
- ✓ Target women as project beneficiaries and ask them what they need and what constraints they see to gender-equitable land administration.
- ✓ Develop implementation guidelines in the local language(s), with women's representation and active participation.
- ✓ Install a gender-sensitive monitoring system, including gender indicators, from the outset.
- ✓ Hire a gender adviser or focal point to work with the implementation team.

Sources: Dalal Alnaggar's presentation on "Developing gender-equitable land survey and technologies in the Arab region", and Kate Dalrymple's presentation on "Gender and technology issues in land administration projects in Southeast Asia and the Pacific" at the May 2011 technical workshop.

✓	Link land administration activities to other projects, such as those for securing women's access to credit and extension services.
✓	Promote the use of gender-sensitive participatory methodologies and tools in regulating access to and use of land and natural resources.
✓	When planning training, communication campaigns and awareness raising, consider context-specific factors such as local languages and customs that may hamper access to the activities for different groups of women and men.
✓	Conduct gender awareness training for women and men, project staff and officials in land administration; offer separate training sessions for women when needed.
✓	Conduct comprehensive public outreach campaigns and education work targeting men and women equally.
✓	Consider language restrictions, illiteracy, and lack of information and transportation, especially for women.
✓	Where national identity documents are used, include a component to assist women and men in obtaining these documents.
✓	Provide information brochures, videos, radio broadcasts, newspaper articles and Web sites that focus on women's situation and use contexts that are familiar and relevant to them, such as by using female figures in illustrations.
✓	Install gender-sensitive evaluation systems and collect best practices and gender-disaggregated data to track long-term impacts on gender equality.

Recording tenure rights

Tenure security can be safeguarded under various forms that make clear the rights of land users and owners: formal titles; clear, long-term rental contracts; reliable lease agreements; or formal recognition of customary and legitimate informal rights, with accessible and effective dispute mechanisms. Many examples show that enhanced tenure security generates individual, household and community benefits by encouraging investments in the improvement of land, homes and neighbourhoods, which improves livelihoods and living standards (UN-HABITAT, 2008: 8). Box 4.3 describes gender-equity issues that arose during the formal recording of traditional rights in South Africa.

BOX 4.3:
Ensuring gender equity in the formalization of traditional rights

Sources: South Africa Department of Land Affairs, 1997: 65 and 68; World Bank/FAO/IFAD, 2009: 456.

A 1997 White Paper on *South African* Land Policy emphasizes the danger that arises from formalized de facto (traditional) rights, under which a land title is usually acquired by one person only, even when more than one person possessed the de facto rights. "People who had previously been equal in their insecurity are now sharply divided when security is vested in one person, generally a male 'head of household'. Experience has shown that it is often women and old people who lose rights in this process" (South Africa Department of Land Affairs, 1997: 65). "For example, Permission to Occupancy certificates have generally been allocated to male household heads. Were they to be upgraded into ownership by the permit holder they would almost invariably be upgraded to the male head of the household. He is often a person who has left to work in the city and may have established a second family there. If he gets upgraded to alienable ownership it creates an incentive to sell the land and could leave the rest of the family vulnerable to dispossession" (South Africa Department of Land Affairs, 1997: 68).

Building on these experiences, the white paper ensures that women have wide-ranging protection through the introduction of new forms of property and procedures for the "upgrading" (formalization) of de facto rights. "Tenure reform provides key opportunities to build protection for the rights of women into the new forms of ownership which are being developed, in particular family-based rights and group ownership systems" (South Africa Department of Land Affairs, 1997: 68).

Land survey, titling and registration[9]

Land survey and registration are core land administration activities for recording land rights in most countries. User-friendly modern technologies for surveying and registering land facilitate quick and efficient field training of both women and men in the day-to-day technical processes of land administration and land tenure governance. Modern technologies are often easy to apply and save costs, thereby supporting gender equity by making technical activities more affordable and accessible to a much wider group of people. Improving gender equity in the technical aspects of land administration is therefore an achievable goal.

For instance, sophisticated global positioning system (GPS) technology is simple to apply in the field through hand-held GPS devices that enable appropriate levels of accuracy. In a land tenure regularization project in Ghana, field investigations in villages were largely carried out by locally trained women, who went house-to-house interviewing people and documenting their land rights and parcel boundaries using laptop computers and hand-held GPS devices. This made the registration process more efficient because data were automatically entered from the field directly into a land information system. As this project found, women can also find it easier to talk about their land to a woman practitioner than to a man (Rebecca Sittie's and Nigel Edmead's presentations at the May 2011 technical workshop; Rebecca Sittie, personal communication).

GPS technology is also an important tool in supporting participatory mapping for gender-equitable land-use planning, as Box 4.4 illustrates with reference to the Philippines.

Parcel mapping based on aerial photographs or satellite images is easier for the whole community to understand – people do not need special training to be able to identify the tree or crop stand that marks the edge of their own piece of land on an aerial photo or satellite image. This method of parcel mapping can therefore be an empowering, accessible and gender-equitable way of recording land rights. Parcel mapping is usually based on local evidence and knowledge about land boundaries, especially from neighbours and other community members, who should be present when an area is being demarcated.

BOX 4.4:
Participatory community mapping using GPS technology

Source: Seema Gaikwad, personal communication.

The *Philippines* Association for Intercultural Development (PAFID) supports indigenous communities in creating participatory 3-D maps and filing claims to their ancestral lands and resources. This process involves the entire community – both women and men receive training in GPS technology and take part in resource mapping, and the resulting maps depend on their combined knowledge of the area being assessed. Young people are particularly receptive learners because they are more confident in exploring and using the new equipment. PAFID has trained more than 148 indigenous people on the use of GPS technology for on-the-ground verification and identification of community reference points. As a result, the Mangyan people of the island of Mindoro have developed a new meaning for GPS in their local language, defining it as Gamit Pang-Sukat – "tool for measurement".

Women participating in the 3-D mapping process identified sources of safe drinking-water and the locations of clinics and schools. During the process, women are on an equal footing with men in the community's decision-making about which data to include in the 3-D maps. Women and men also have equal opportunities to use the maps. Women participate from the initial construction of the model to the coding and interpretation of the data.

In 2003, PAFID was asked to export its participatory 3-D mapping approach to other countries and to carry out training sessions in *India and Myanmar*.

Making land registration and titling more accessible and gender-equitable

Land administration institutions in many developing countries lack office and staff capacity. They may be remote from rural communities and have limited budgets and resources for transportation to the field. In turn, rural people's access to land administration offices is constrained by a lack of transport, good roads and money, particularly for women who in many societies have less opportunity to travel than men. One way of making land registration more accessible to women is through the decentralization of land administration institutions, as Box 4.5 from Ghana describes.

BOX 4.5:
Making land registration more accessible to women

Source: Rebecca Sittie's presentation on "Gender issues in managing and implementing land registration in Ghana" at the May 2011 technical workshop.

In *Ghana*, the number of documents registered by women in their own names in the deeds registration system increased substantially between 2005, when there were only two land registries in the whole country, and 2006, when land administration was decentralized and more registries were opened outside the capital (see the figure below). Decentralization was accompanied by a public awareness campaign, informing women and men about the opening of new deeds registry offices where they could register their documents.

No. of Documents Registered by Women 2005 - 2010

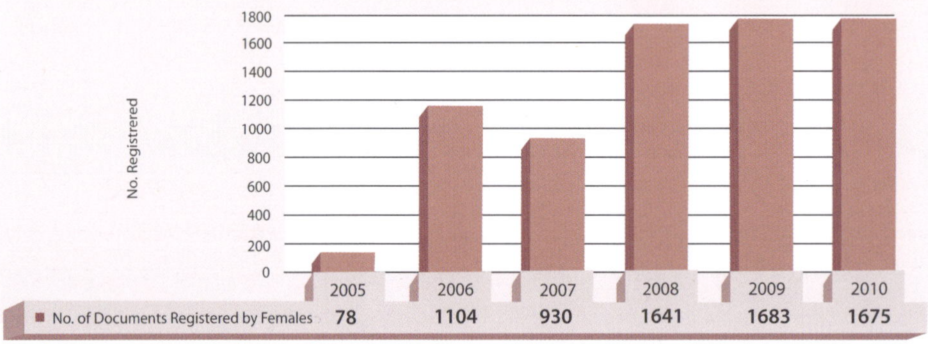

The charges and fees for land registration services also have an impact on accessibility to women, as the examples in Box 4.6 show.

BOX 4.6:
Acting on registration fees to support gender equity in land administration

Sources: Seema Gaikwad's presentation on "Asian experiences of civil society engagement with land policy processes", and Rebecca Sittie's presentation on "Gender issues in managing and implementing land registration in Ghana" at the May 2011 technical workshop; Undeland, 2012.

In *Nepal*, the government adopted a directive that waives part of the land registration fee when land is registered in a woman's name. The exemption started at 10 percent in 2006, increasing to 20 percent in 2007 and 25 percent in 2009. (An exemption of 50 percent is now being demanded). As a result, the amount of land registered in women's names more than tripled, although it is difficult to know whether this increase has led to more secure women's land rights in practice.

In two projects in *Ghana*, the costs of registration were substantially reduced to encourage people to register their land. The government's World Bank-funded Land Administration Project reduced the cost of title registration in urban areas to about a quarter of its normal cost. In the Rural Systematic Titling project (described in Box 5.2), the government gave exemptions on stamp duty and the Millennium Challenge Account covered the total costs of documentation, survey and mapping, registration and legal fees. Although questions arise regarding the sustainability of this kind of support, it is likely to be an important tool in boosting the gender equity of land registration and land administration as systematic titling is rolled out into Ghana's rural areas.

In *Kyrgyzstan*, the cost of registering property and land has been identified as a major deterrent to women, despite the country's exemplary system of registering all co-owners.

CHECKLIST 4.2:
Improving gender equity in surveying and land registration

✓	Prior to land registration, conduct an assessment of existing rights, focusing on women's rights (registered and unregistered) and inheritance, marriage and divorce patterns.
✓	Develop implementation guidelines and manuals for field staff, and ensure women's participation.
✓	Install a gender-sensitive monitoring system, including gender indicators.
✓	Target female-headed households for land registration and ask them what they need and where they see constraints in gender-equitable land titling.
✓	Promote the participation of women operators in the registration process.
✓	Promote joint titling and the registration of women as the exclusive owners of their land.
✓	Design the land information system and databases to include steps that ensure the law and regulations regarding gender equity are implemented.
✓	Support parcel mapping based on local evidence and knowledge about land boundaries, ensuring the presence of women (including from male-headed households), neighbours and other community members in demarcation processes.
✓	Decentralize land registries to make them more accessible to rural communities.
✓	Combine land-titling activities with economic development projects such as agricultural extension, small business development or microcredit.
✓	Keep fees low to facilitate access for women and other vulnerable groups.
✓	Carry out public awareness campaign that target women, informing them of the importance of land rights and registration, where rights can be registered and under which conditions – service fees, need of proof, etc.
✓	Consider illiteracy and lack of information and transportation.
✓	Provide information brochures, videos, radio broadcasts, newspaper and Web sites for women, which may need to be arranged differently from those for men.
✓	Verify that titles are correct and that all property owners have been legally titled.
✓	Ensure that titles are delivered simultaneously to the women and men owners of property.
✓	Facilitate the appeal process for women by establishing links with lawyers and paralegals who can offer specific assistance.
✓	Install gender-sensitive evaluation systems, and collect best practices and gender-disaggregated data to track long-term impacts of gender equality.

Source: Dalal Alnaggar's presentation on "Developing gender-equitable land survey and technologies in the Arab region" at the May 2011 technical workshop.

Alternatives to land titling

Attempts to formalize informal rights and customary tenure have often focused on granting individual freehold titles. Land titling has been popular with many international agencies and also with beneficiaries. However, care must be taken to ensure that formalization does not result in women losing the secondary rights – access, use, collection or cultivation rights – that they enjoyed under customary law (Box 4.3). By undermining pre-existing land rights, titling can increase insecurity.

No single form of tenure can meet the different needs of all social groups. When a range of land tenure options are recognized, both women and men from all social groups are more able to meet their changing needs over time. Legal recognition for different forms of tenure can also promote the development of dynamic land markets (UN-HABITAT, 2008a: 10). In most countries, the different land rights and tenure types available form a continuum (Figure 4.1). At one end of this continuum are informal or illegal shared forms of tenure, such as de facto recognition of an informal or illegal settlement, or political protection against forced eviction. In the middle are occupancy rights, shared user rights, collective adverse possession, special concessions and customary communal tenure, which can be either informal or formal. At the other end of the continuum are co-ownership and registered leases, conventionally referred to as "formal" tenure types (UN-HABITAT, 2007b: 24).

FIGURE 4.1: **Continuum of land rights**

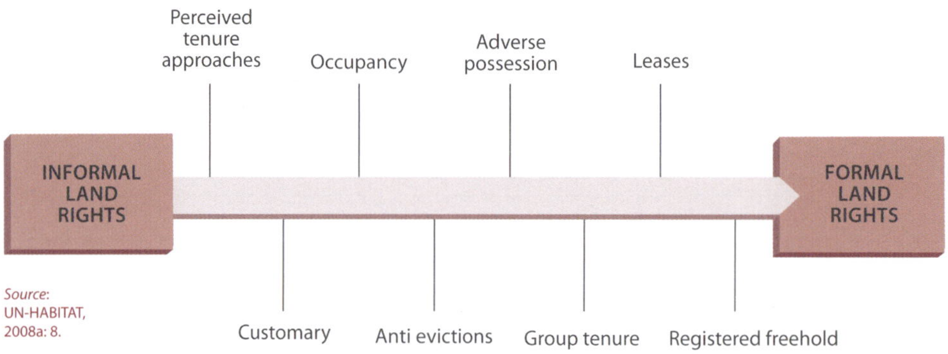

Source: UN-HABITAT, 2008a: 8.

Alternatives to titling (Box 4.7) can provide tenure security, but must be selected according to the context – rural, peri-urban, urban, informal settlement, etc. The extent to which these forms of tenure are gender-equitable is very variable, as is the tenure security attached to each, which depends on several factors. For example, compared with freeholds, leases – particularly tenancy contracts – are cheaper, can be delivered faster, are more flexible, can be made even when landownership is under dispute, can be upgraded as needed, and require cheaper and simpler technological systems. Measures such as joint registration of leases or their recording in the names of both spouses may be necessary, to avoid the exclusion of women (Augustinus and Benschop, 2003: 4).

Informal tenure types such as the de facto recognition of occupation – through political patronage, proof of payment of utility bills, oral evidence, informally recognized customary rights, perceived secure tenure, etc. – are the most common types in slums and informal settlements. The security of tenure that they provide depends on local circumstances and on whether they are accompanied by other forms of protection against forced eviction (Augustinus and Benschop, 2003: 4).They can provide the foundation for incremental approaches to tenure improvement, in which tenure rights are gradually formalized or upgraded over time. An incremental approach can increase short- to medium-term security while the authorities develop more comprehensive and locally sensitive long-term alternatives (UN-HABITAT, 2008a: 10–11 and 20).

BOX 4.7: Alternatives to titling

Security and services in Colombia

Formal tenure systems are often considered essential in ensuring access to public services. In Colombia, however, legislation allows all citizens to obtain services such as water, sanitation, electricity, storm drainage, garbage collection, telephone and gas as long as they can prove that they live in their homes and can pay for the services provided. A range of intermediate tenure systems – such as declarations of possession, buying and selling of rights for future use, and communal tenancy – provide stepping-stones towards strengthened rights and protection from eviction, enabling poor households to obtain relatively secure, affordable housing regardless of their tenure status.

Temporary occupation licences in Kenya

Nairobi City Council introduced temporary occupation licences (TOLs) to promote the efficient use of idle public land in strategic locations, such as street intersections, road reserves in high-density neighbourhoods and open land on the urban fringes. Under a TOL, the local authority allocates land for a productive use while retaining long-term control; the licence is renewable annually and the licensee is allowed to build semi-permanent structures. Licensees pay annual land rents, which add to municipal revenues. Although intended for commercial land income-generating purposes, TOLs are also sometimes used for residential purposes and could be replicated elsewhere.

Temporary land rental in Thailand

Low-income communities in Bangkok have evolved a practical arrangement with landowners that enables them to live in areas with affordable access to livelihood opportunities. Poor people look out for landowners who are waiting for their land to gain in value before developing it, and offer to rent the land on a short- to medium-term lease, paying what they can afford. Landowners favour such rentals as protection against third-party invasion of their land. In recent years, communities and authorities have been exploring ways of extending basic services to these temporary settlements. As urban growth brings livelihood opportunities to other locations, poor people can negotiate similar arrangements with landowners in these new areas.

Sources: Aristazabal and Gomez, 2004; Yahya, 2002; Farfan, 2002.

Gender issues in designing and setting up land information systems and databases[10]

Although gender issues are not usually built into land information systems and databases, the design of any land information system must fit into the relevant policy and legal framework, including for gender issues. For instance, if joint titling is mandated by law, the system should be designed to support this. The computerization of paper-based land information systems provides opportunities for incorporating gender issues, such as by introducing systems for recording the gender of landowners, to allow gender disaggregation and analysis of data. Data collection forms and registration documents should include enough space for the names of all landowners and people with interests in the land.

Computerized databases and systems can easily be set up to record spousal rights, joint land ownership, and family members' interests in land, and to require the submission of legally mandated certificates, other documents or pieces of information (e.g. marriage certificates) before proceeding to the next step in the data entry process. In Uganda, for example, land transfers require the submission of Land Form 41 – "Consent by spouse(s) to transaction in land" – so the land information system is designed to ensure that this form is scanned before the next step in the formal transfer of title can take place.

To make land administration more gender-equitable, women practitioners should participate in the design and use of tools and technologies, and training in the administration and operation of land information systems and databases should be designed to meet the educational needs of both women and men. Tools should be evaluated from a gender perspective; the Global Land Tool Network's (GLTN's) gender evaluation criteria, developed through the Gendering Land Tool Development Mechanism, is an excellent example of good practice here (UN-HABITAT, 2007a; 2009; 2011).

CHECKLIST 4.3: **Ensuring gender-equitable land information systems**

Source: FAO, 2009c: 115.

✓	Use low-cost technologies that are accessible to wider groups of people.
✓	Ensure women's participation in the development of questionnaires and surveys.
✓	Collect gender-disaggregated data.
✓	When collecting data, interview women separately from men, to get their side of the story.
✓	Use participatory community mapping for land-use planning and resource assessment, involving women and youth separately from men and older adults, and ensuring the presence of women in all working groups.
✓	Demystify technology and employ more women practitioners in fieldwork, to help demonstrate that land administration activities are gender-equitable.
✓	Ensure that data collection forms and registration documents allow space for the names of all landowners and people with interests in the land, and for recording gender.
✓	Evaluate tools from a gender perspective (e.g. through GLTN's gender evaluation criteria).

REMEMBER!
Even when technology is gender-neutral, its usage may not be.

Valuation[11]

As many women have fewer financial resources than men, it is particularly important that women's properties are valuated fairly. Fair valuation takes into account the two opposing purposes of property valuations: i) for calculating property tax; and ii) for calculating compensation to owners in cases of expropriation.

CHECKLIST 4.4: **Supporting gender-equitable valuation**

Source: Nigel Edmead's presentation on "Gender issues in designing and setting up land information systems and databases: experiences from Ghana, Zambia and Uganda" at the May 2011 technical workshop.

✓	Ensure that valuation systems take into account non-market values, such as social, cultural, religious, spiritual and environmental values.
✓	Assess the influence of social factors – social relations, prestige, fraternity – on negotiations to establish the price/value of property, especially for vulnerable groups such as widows and women heads of household who have weak political bargaining power.
✓	Keep fees to a minimum.
✓	Explain to landowners how valuations are decided, especially to women who often have lower literacy levels and lack information.
✓	Prevent corruption by making valuation information and analyses available to the public.

Taxation[12]

Tax policies have impacts on poverty, income distribution and gender equity. Care must be taken to ensure that tax policies do not exacerbate discrimination against men or women, especially vulnerable groups such as female-headed households. Heavy taxation of property (co)ownership can encourage people to avoid formal registration and remain under less secure forms of tenure. As a general rule, annual land tax should not exceed one percent of the property value, and land acquisition tax should not exceed three percent.

CHECKLIST 4.5: **Supporting gender-equitable taxation**

✓	Ensure that women and men are treated equally in the tax code and regulations.
✓	Address the differential impacts that land tax could have on women and men – their shares in landownership, land use and livelihoods, and land transfer practices.
✓	Ensure that the political debate regarding tax policy is fully informed about the social impacts of tax policy options, especially on gender equality.
✓	Identify a tax base that is easy to administer and assess, but also fair and justifiable.
✓	Ensure that taxes are based on appropriate values.
✓	Consider granting tax exemptions and relief to vulnerable groups, such as poor women.
✓	Assess how acceptable the tax is to tax-payers, and explain to them the need for and use of the taxes raised. Tax-payers may be involved in setting tax rates and deciding how the tax revenue should be used.
✓	Assess and collect taxes in ways that do not discriminate against women or men – taking into account issues such as illiteracy and lack of information or transport – and that can be combined with other activities such as community meetings, markets or training sessions.
✓	Ensure that tax revenues are used in ways that benefit both women and men, by involving both genders in needs analysis, such as through participatory budget planning at the municipal level.
✓	Conduct training on filling in tax forms and the amounts to be paid in different situations – property transactions, real estate acquisitions, including through inheritance or gifts, etc.
✓	Ensure transparency by making assessments public.
✓	Provide procedures that give women and men the opportunity to appeal for review of their tax assessments.

Source: FAO *et al.*, 2007.

Land-use planning[13]

Land-use planning affects the ways in which people can use the land they have rights to. In many countries, women and men use land in different ways, so gender-equitable governance of land tenure requires gender-sensitive land-use planning.

Participatory land-use planning at the grassroots is a good way of achieving this, as the example of community mapping from the Philippines described in Box 4.4 shows. Box 4.8 illustrates other good practices from Ethiopia and Mali. The seven-step approach described in the Ethiopian case can be replicated in other contexts and countries.

BOX 4.8:
Gender-balanced participatory land-use planning

Source: Wehrman, 2011: 156–157.

In *Ethiopia*, the German Agency for International Cooperation's (GIZ's) project on Land-Use Planning and Natural Resource Management in Oromia Region incorporated gender mainstreaming into its design. Recruitment procedures promoted a gender balance of women and men staff – when this was not possible, women consultants and interns were hired instead. Project and counterpart staff were trained on integrating gender issues into all activities. Participatory land-use planning and participatory rural appraisal activities involved women and men from the target populations, and their timing and location was planned to avoid disadvantages for either women or men. Gender analysis tools were used to collect information about the access to and control over resources, activities and workloads of all household members and the participation of women in decision-making at different levels. Project surveys established gender-disaggregated data, which were used to analyse existing land uses in project districts and communities.

Women now participate fully in participatory land-use planning in Oromia and take part in public and family decision-making. Economic activities initiated by the project have continued through traditional women's groups, with increasing numbers of women taking part.

The separate transact maps carried out by women and men in *Mali* demonstrated that women provide valuable contributions to the planning and management of land and other natural resources when they are involved in participatory land-use planning. The women's transact focuses on land-use problems related to the women's vegetable gardens, rice fields and access to water. It covers a far smaller area than the men's transact, but provides far more detail about this part of the village, which is primarily used by women. The men's transact focuses on the more remote areas that are more relevant to their work. Hence, the two transacts complement each other and both contribute equally to the overall picture of the local resource situation.

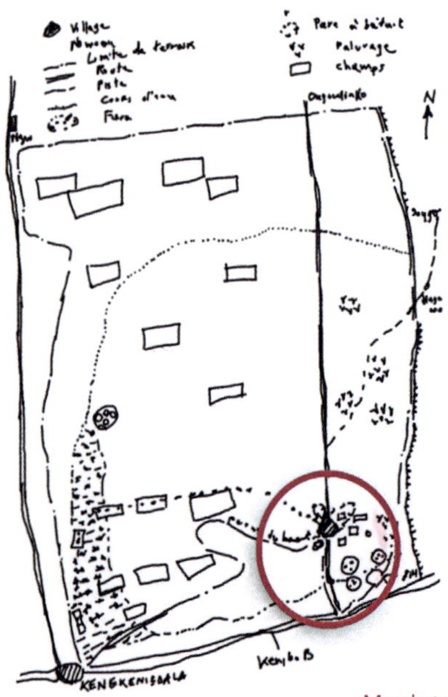

Men's transact

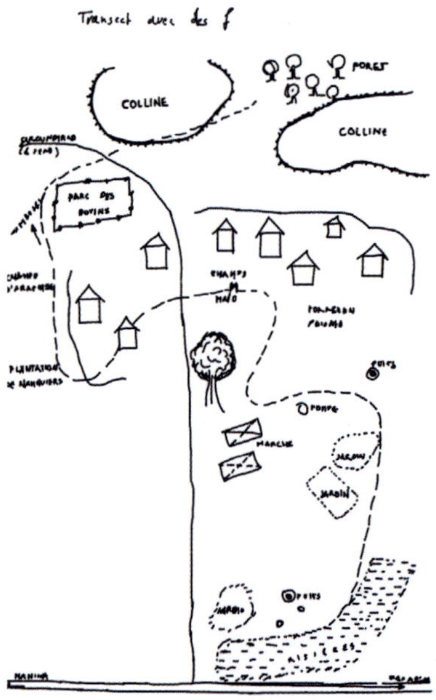

Women's transact

CHECKLIST 4.6: Supporting gender-equitable land-use planning

✓	Involve women in clarifying the need and/or demand for land-use planning (including the updating of existing plans) as they may see demands and needs that men do not see.
✓	Analyse the situation on the ground: existing land rights – registered and unregistered – especially of women; different uses of and interests in land; etc.
✓	When identifying stakeholders to participate in land-use planning workshops ensure the inclusion of women experts, members of women's associations and women who represent different social groups.
✓	Address these women directly when sensitizing or informing stakeholders.
✓	Ensure that women are contacted when entering into dialogue with participants (land users).
✓	Ensure women's equal representation in data collection interviews, group discussions and workshops. Times and locations of meetings must be adapted to women's schedules and accessibility.
✓	Contact women separately from men to ensure that they address the issues that are important to them.
✓	Involve women in data collection -- analyses of situations, needs, potentials, etc. – as they often hold important information that is not available to, or valued by, men.
✓	Create awareness on gender issues in land-use planning.
✓	Identify feasible measures for alleviating gender inequalities.
✓	When the plan is being formulated (often by experts), consider women's inputs as being as important as those of men.
✓	With gender-sensitized experts, discuss the possible impacts of measures on both women and men.
✓	Involve women in decision-making.
✓	Ensure that the plan responds to the needs and expectations of women as much as to those of men.
✓	Ensure that the priority activities implemented first benefit women and men equally.
✓	Base implementation monitoring of the land-use plan on gender-disaggregated data.

Land consolidation and readjustment[14]

Consolidation of land parcels, whether of housing or agricultural land, and reallocation of rights within new boundaries can have major impacts on the existing landownership and use rights of women and men. It is therefore vital to consider these existing rights and the land-use needs of affected women and men. Participatory approaches to designing land consolidation strategies and projects are particularly helpful.

CHECKLIST 4.7: Supporting gender-equitable land consolidation

✓	Analyse existing boundaries and the legal status of parcels, focusing on women's rights (registered and unregistered).
✓	Identify what the community needs and wants, consulting women and youth separately from men and older adults.
✓	Ensure that national policies relating to land consolidation guarantee the rights of both women and men and support the inheritance rights of women and girls.
✓	Provide gender-sensitivity training for officials working on consolidation issues.

✓	Share a draft concept plan – stating the aims, area, scope and schedule of the consolidation – with women and men of the affected community, and make sure that all the people involved have access to this information (through newspapers, radio, the Internet).
✓	Collect data on property, land use and management and other socio-economic indicators (literacy, level of education, sources of income, etc.) in the community, disaggregated by gender.
✓	Include community women and men in the land consolidation board. Facilitate women's participation in meetings, and organize meetings for women only.
✓	Ensure transparency regarding the valuation of parcels.
✓	Include in planning and decision-making all the women and men whose parcels are affected, regardless of who the official owners are.
✓	Share a detailed plan of the land consolidation – showing the new parcel layout, location of new roads and other public facilities – with women and men, and identify the roads and facilities that will be removed.
✓	Incorporate comments into the plan and give both women and men the opportunity to raise objections regarding boundaries, ownership, valuations, etc.
✓	Organize a participatory review of the consolidation options.
✓	Inform the community about the approval, survey of new boundaries on the ground, final update of cadastral maps, and issuing and registration of new titles.
✓	Establish fair and transparent compensation schemes that do not discriminate against women and that involve payments to both spouses and all landowners.
✓	Promote legal assistance for women advocating women's rights.

Land redistribution and restitution[15]

In the privatization process that followed the collapse of communism in Eastern Europe, land restitution was widely applied to return land to its original owners whose ownership rights had been removed under laws created by former communist governments (Osskó, 2002). (To a greater or lesser extent, these governments had followed the lead of the Soviet Union, which made all land the property of the State.) Although the gender-differentiated impacts of the restitution process have not been widely studied, there are indications of unequal treatment of women and men in the implementation of new land laws, and evidence of the revival of local-level patriarchal norms and practices regarding women's rights to land (Box 4.9), even where the formal privatization rules do not disfavour women.

Box 4.9:
The impacts of land redistribution in Eastern Europe on women

Sources: World Bank, 2005: 8; Giovarelli and Duncan, 1999.

In *Albania*, land has been redistributed to farming families and the former members of cooperatives and State farms, according to the number of people in the family. However, land has been titled and registered in the name of the head of household only.

In *Azerbaijan*, the 1996 Land Reform Law required that land previously held in collective ownership be divided and distributed to rural residents. Under the World Bank's Farm Privatization Project, all the household members aged at least 16 or 18 years received shares, and their names were listed on the land certificates, although only the names of household heads were entered in the land registry. This redistributed land was held in common ownership. Azerbaijan has retained very old traditions that do not encourage this type of distribution; women do not receive land under the country's customary marital property and inheritance laws, even though under the 1996 law, women have rights to farmland. As a result, on several of the redistributed farms, women have received smaller plots than men.

Gender-equity is particularly critical in the land restitution activities that follow humanitarian, conflict and natural disasters (UNHCS, 1999). Women often face enormous barriers in access to land after conflict and natural disasters, and when land tenure governance in general is severely challenged, gender equity may not seem an immediate priority. In Colombia, however, the Victims and Land Restitution Law of 2011 aims to redress the losses suffered by victims of the protracted internal armed conflict. The new law adopts a gender-sensitive approach and takes into account special characteristics based on age, sexual orientation and disablement. The Regulations for Women in Restitution Processes prioritize women in land restitution procedures and the formal titling of restituted possessions. Land deeds are issued and registered jointly to spouses or permanent partners, even if the spouse or partner is not present during the administrative procedure (Meertens, 2012).

Mainstreaming gender concerns in disaster mitigation and preparedness activities also increases the likelihood that women and men will develop the capacity to anticipate and cope with the effects of natural disasters (World Bank/FAO/IFAD, 2009b: 451).

The Pinheiro Principles provide guidance on housing and property restitution issues for refugees and internally displaced persons and are relevant to gender-equitable governance of land tenure because they uphold core values of non-discrimination and gender sensitivity. In particular, Principle 14 calls for adequate consultation and participation in decision-making, including women's representation and inclusion in restitution decision-making processes, and for ensuring that women have the means and information necessary to participate effectively (FAO et al., 2007). The principles are relevant to land restitution activities in all post-disaster settings.

CHECKLIST 4.8:
Supporting gender-equitable land restitution

✓	Examine the existing rights, taking into account the land rights for which there is no clear evidence.
✓	Ensure that national policies relating to land, housing and property restitution guarantee the rights of both women and men and support the inheritance rights of women and girls.
✓	Determine a policy for restitution, including which property rights will be restored to which claimants, and the form of restitution used, such as recovery of the property or payment of compensation.
✓	Provide training in gender sensitivity for officials working on restitution issues.
✓	Provide timely information about how and when to make restitution claims, via newspapers, the radio, the Internet and public meetings.
✓	Design programmes to assist women and girls in making restitution claims, for example by providing simple claim forms and offering gender-sensitive assistance in their completion.
✓	Provide special resources to households headed by single women so they can avail themselves of their rights in emergency settings.
✓	Mainstream gender sensitivity throughout programmes and procedures, so women and men enjoy equal treatment in the restitution process, up to and including the joint conferral of any housing, land or property rights that are confirmed.
✓	Monitor women's housing and property restitution rights to ensure that women refugees exercise their rights voluntarily, and are neither coerced into returning to their homes nor prevented from doing so in safety and dignity. Implement reparation diligently and promptly. Ensure that legal assistance for women and men is available at local levels, and consider creating mobile legal teams.

Supporting gender-equitable land restitution

- ✓ Support specific disaster mitigation measures to reduce the vulnerability of female-headed, displaced and poor households.

REMEMBER!
Women and men may each focus on different aspects of the restitution process, and efforts should be made to determine their different views and how these can be taken into account and facilitated throughout the entire restitution process.

Redistributive reforms[16]

While land consolidations and readjustments are undertaken to improve the layout and structure of land parcels, redistributive reform programmes are much more broad-ranging and involve the removal of land rights from some landowners and users and their reallocation to new beneficiaries, generally countrywide. In many developing countries, redistributive reforms are an important part of policies for increasing access to land, and often target landless, poor and marginalized beneficiaries, including women and female-headed households.

Communication issues are particularly important in ensuring gender-equitable redistributions, as discussed with respect to the Government of Sindh's Land Distribution Programme in Pakistan in Box 5.3.

CHECKLIST 4.9: **Supporting gender-equitable redistributive reforms**

- ✓ Include women and men in the national debate about land redistribution objectives.
- ✓ Facilitate women's participation in technical working groups and commissions engaged in fact finding and the drafting of recommendations for development of a land redistribution policy.
- ✓ Encourage cooperation from the line ministries implementing the land reform process in providing joint support to beneficiaries, especially female-headed households.
- ✓ Limit the contributions that beneficiaries are required to make in cash, in kind or by loan, and allow beneficiaries to make group applications, to ensure inclusion of poor and vulnerable groups and promote poverty alleviation.
- ✓ Promote gender equality in laws and regulations for land allocation.
- ✓ Conduct a land needs assessment – farming, subsistence, etc. – and allocate land parcels accordingly, based on their fertility, size and location.
- ✓ Help vulnerable groups such as the landless and women to identify, appraise and negotiate land purchases and to participate in the planning of farm development.
- ✓ Ensure transparency to prevent influential people from steering beneficiary selection.
- ✓ Define clearly the property rights allocated to resettled people. Rights and responsibilities must be well understood and accepted by those who have to abide by them, strictly enforced by the State and perceived as being secure.
- ✓ Establish mechanisms for preventing and resolving conflicts regarding boundaries or the use of natural resources that result from the land redistribution, to facilitate the resettled people's smooth integration into the surrounding community, mediation mechanisms, agrarian courts, etc.
- ✓ Choose beneficiaries with experience as tenants or workers; support those without this experience by providing agricultural training in appropriate and sustainable land use.
- ✓ Facilitate access to input and output markets, credit and advisory services for both women and men.
- ✓ Have a sound monitoring system for identifying changes in outcomes, refining programmes, and planning mitigation measures for negative events and impacts.

Compensation[17]

Compensation is generally required for expropriations carried out in the public interest, and sometimes also for people who lose tenure rights during land restitution, consolidation and redistributive processes. The question of who receives the compensation on jointly owned land can be very complicated when landownership is under dispute. Laws relating to compensation should therefore require investigation of who will suffer losses from any compulsory land acquisition or expropriation, paying special attention to the land-use rights of women and children who may not be formal landowners, and to the rights of tenants. Compensation payment mechanisms should be designed to ensure joint family decision-making about use of the compensation funds (FAO, 2008a: 33–34).

CHECKLIST 4.10: **Supporting gender-equitable compensation**

- ✓ Check who owns the land or has rights to it, making sure that all owners are duly notified regardless of whether or not they live on the land that is being expropriated.
- ✓ Check that all the title holders of jointly owned land receive equitable shares of compensation.
- ✓ Ensure that fair compensation is paid to the landowner(s). Where possible, consider land-for-land compensation for vulnerable groups without other income.
- ✓ Compensation payment mechanisms should ensure joint family decision-making on use of the funds.

REMEMBER!
To ensure that women benefit from compensation. It is crucial that their tenure rights are recognized and recorded.

Notes

[7] Dalal Alnaggar, Marianna Bicchieri, Kate Dalrymple, Nigel Edmead, Neil Pullar and Rebecca Sittie provided substantial inputs to this module during the May 2011 technical workshop at FAO Headquarters, including during discussions of the working group on technical issues.
[8] World Bank/FAO/IFAD. The Gender and Agriculture Sourcebook, 2009. http://siteresources.worldbank.org/INTGENAGRLIVSOUBOOK/Resources/CompleteBook.pdf
[9] Ibid.
[10] Nigel Edmead's presentation on "Gender issues in designing and setting up land information systems and databases: experiences from Ghana, Zambia and Uganda" at the May 2011 technical workshop.
[11] Mends, 2006: 5ff; Ambaye, 2009: 13 ff.
[12] www.fig.net/commission9/wgroups/sg9_1_1_11_14.htm; Barnett and Grown, 2004.
[13] Wehrman, 2011.
[14] FAO, 2003b: 22.
[15] Grover and Flores-Bórquez, 2004; Osskó, 2002.
[16] Binswanger-Mkhize, Bourguignon and van den Brink, 2009: 14ff.
[17] Binswanger-Mkhize, Bourguignon and van den Brink, 2009: 23ff; Ambaye, 2009: 18ff.

Summary of key messages on technical issues

Before starting any land administration activities the situation on the ground should be analysed, including family structure, women's decision-making power, inheritance, marital and divorce patterns, representation of women in local affairs, different uses of and interests in land, and existing boundaries and land rights (registered and unregistered), particularly women's rights.

Mechanisms should enable the registration of women's specific land rights, such as use, secondary or joint-ownership rights. Women should be able to exercise the right to exclusive ownership of property – individually or as groups.

Participatory methodologies and tools that enable women to take part should be promoted. Separate meetings for women should be organized to ensure that women address the issues that are important to them, which for cultural reasons they may not do in the presence of men.

Women representing different social groups should be identified as key stakeholders, and their active participation and involvement in decision-making processes facilitated. Illiteracy, lack of information and transport should be taken into account, and the location and timing of meetings adapted to women's schedules and mobility.

Attention should be paid to gender equity issues in the design of all new land information systems and databases.

Legal and procedural reforms may be needed to support the use of technologies and systems that facilitate improved gender equity in land administration.

Land tools should be evaluated from a gender perspective.

Gender equity issues should be mainstreamed into the design, implementation and monitoring of all land administration projects and activities. There is need for gender-equitable participation in technical aspects of land tenure governance – land administration activities, technologies and systems – to accommodate the interests of all people – women and men.

Staff, government officials and communities should be trained in gender sensitivity and public outreach campaigns should be conducted, to raise awareness for gender-equitable land administration activities.

The procedures and results of land administration activities must be easy to understand, available to the public and low-cost, to avoid corruption and exclusion.

The differential impacts of measures on women and men should be considered to ensure that land administration activities benefit women and men equally.

Suggested monitoring indicators for technical issues

- % of women and men working in the technical professions of land administration (surveying, valuation, geodetics, GPS, etc.)
- number of women enrolled in and graduating from technical courses in higher education and scientific subjects at secondary school
- % of women and men receiving advocacy and legal literacy training focusing on land
- number of gender-sensitive technologies used in land administration activities
- % of women and men attending training on use of new land-related technologies

Ambaye, D. W. 2009. Land valuation for expropriation in Ethiopia: Valuation methods and adequacy of compensation. Paper presented at the 7th FIG Regional Conference on Spatial Data Serving People: Land Governance and the Environment – Building the Capacity, Hanoi, 19–22 October 2009.

Aristazabal, N. & Gomez, A. 2004. Improving security without titles in Bogotá. *Habitat International,* 28(2): 245–258.

Augustinus, C. & Benschop, M. 2003: Security of tenure – best practices. Nairobi, UN-HABITAT. http://www.unhabitat.org/downloads/docs/10784_1_594339.pdf

Barnett, K. & Grown, C. 2004. Gender impacts of government revenue collection: The case of taxation. *Economic Paper No. 62.* London, Commonwealth Secretariat.

Binswanger-Mkhize, H. P., Bourguignon, C. & van den Brink, R. 2009. *Agricultural land redistribution: toward greater consensus.* Washington, DC, World Bank.

FAO. 2003b. *The design of land consolidation pilot projects in Central and Eastern Europe. Land Tenure Studies No. 6.* Rome. ftp://ftp.fao.org/docrep/fao/006/y4954e/y4954e00.pdf

FAO. 2008a. *Compulsory acquisition of land and compensation. Land Tenure Studies No. 10.* Rome. ftp://ftp.fao.org/docrep/fao/011/i0506e/i0506e00.pdf

FAO. 2009c. "We all inherited this earth", inheritance rights of widows and children in the context of AIDS: legal overview. Rome. (final draft)

FAO. 2012b. Voluntary Guidelines on the Responsible Governance of Tenure of Land, Fisheries and Forests in the Context of National Food Security.Rome. http://www.fao.org/docrep/016/i2801e/i2801e.pdf

FAO, Norwegian Refugee Council, UNHCR, UN-HABITAT, OHCHR & OCHA. 2007. *Handbook on housing and property restitution for refugees and displaced persons – implementing the "Pinheiro Principles".* www.ohchr.org/Documents/Publications/pinheiro_principles.pdf

FAO. 2012a. Improving Gender Equality in Territorial Issues (IGETI).Land and Water Division Working Paper No. 3. Rome. www.fao.org/docrep/016/me282e/me282e.pdf

Farfan, F. 2002. *In* G. Payne, ed. *Land rights and innovation: improving tenure security for the urban poor.* London, ITDG Publishing.

Giovarelli, R. & Duncan, J. 1999. Women and land in Eastern Europe and Central Asia. Washington, DC, USAID. http://pdf.usaid.gov/pdf_docs/pnacl400.pdf

Giovarelli, R. & Lastarria-Cornhiel, S. 2006. *Study on women and property rights: project best practices.* Washington, DC, USAID. http://pdf.usaid.gov/pdf_docs/pnadj420.pdf

Grover, R. & Flores-Bórquez, M. 2004. *Restitution and land markets.* Paper presented at the FIG Working Week, Athens, 22–27 May 2004.

Meertens, D. 2012. *Colombia's law on victims and restitution: a challenge for gender-sensitive transitional justice.* Paper presented at the Expert Group Meeting on Good Practices in Realizing Women's Rights to Productive Resources, with a Focus on Land, UN Women in collaboration with the United Nations Office of the High Commissioner for Human Rights, Geneva, 25–27 June 2012.

Osskó, A. 2002. Land restitution and compensation procedures in Central Eastern Europe. Paper Presented at the FIG Commission 7 Annual Meeting, Pretoria, 4–9 November 2002.

South Africa Department of Land Affairs.1997. White Paper on South African Land Policy. Pretoria.

Teklu, A. 2005. Research report 4 – land registration and women's land rights in Amhara Region, Ethiopia. London, IIED.

Undeland, A. 2012.Lessons learned and good practices in realizing women's rights to productive resources with a focus on land: Kyrgyzstan and Tajikistan. Paper prepared for the Expert Group Meeting on Good Practices in Realizing Women's Rights to Productive Resources, with a Focus on Land, organized by UN Women in collaboration with the United Nations Office of the High Commissioner for Human Rights, Geneva, 25–27 June 2012.

UNCHS. 1999. Women's rights to land, housing and property in post-conflict situations and during reconstruction: A global overview. Nairobi, United Nation Centre for Human Settlements.www.unhabitat.org/downloads/docs/1504_59744_Land.pdf2.pdf

UN-HABITAT. 2007. Gendering land tool development. Nairobi, GLTN Secretariat.

UN-HABITAT. 2007b. Policy makers guide to women's land, property and housing rights across the world. Nairobi.

UN-HABITAT. 2008a. Secure Land Rights for All.Nairobi.

UN-HABITAT. 2009. Gender evaluation criteria for large-scale land tools. Nairobi.

UN-HABITAT. 2011. Designing and evaluating land tools with a gender perspective: training package for land professionals. Nairobi.

Wehrmann, B. 2003. Women's access to and control over land in African and South Asian countries. *Landnutzung und Landentwicklung* [Land Use and Development], 44: 83–89.

Wehrman, B. 2011. *Land use planning. Concept, tools and applications.* Eschborn, Germany, GIZ. www2.gtz.de/dokumente/bib-2011/giz2011-0041en-land-use-planning.pdf

World Bank. 2005. Gender issues and best practices in land administration projects: A synthesis report. Washington, DC. http://siteresources.worldbank.org/intard/resources/gender_land_fulltxt.pdf

World Bank/FAO/IFAD. 2009. Gender in agriculture sourcebook. Washington, DC and Rome.

World Bank/FAO/IFAD. 2009a. Module 4: Gender issues in land policy and administration. *In* Gender in agriculture sourcebook. Washington, DC and Rome.

World Bank/FAO/IFAD. 2009b. Module 10: Gender and natural resources management. In *Gender in agriculture sourcebook.* Washington, DC and Rome.

MODULE

5

Getting the message across

LIST *of* ILLUSTRATIONS

Boxes

Checklists

Tables and figures

MODULE 5: **Getting the message across**

Communication and awareness-raising strategies and methods for gender-equitable governance of land tenure

Introduction[18]

Communication and awareness raising are crucial to changing values and attitudes, including those about gender and land tenure governance. Communication can mobilize people for development actions, such as claiming their land rights, while helping to build consensus on problems and their possible solutions through the participation of all stakeholders (FAO/GTZ, 2006).

One of the ten core principles of implementation of the Guidelines – transparency – defines how communication and awareness raising can support responsible gender-equitable land tenure governance. The Guidelines also stress the need for gender-sensitive and gender-targeted messages (Box 5.1).

BOX 5.1: **Getting the message across and the Guidelines**

3.B.8 Transparency: clearly defining and widely publicizing policies, laws and procedures in applicable languages, and widely publicizing decisions in applicable languages and in formats accessible to all.

8.9 States should allocate tenure rights and delegate tenure governance in transparent, participatory ways, using simple procedures that are clear, accessible and understandable to all, especially to indigenous peoples and other communities with customary tenure systems. Information in applicable languages should be provided to all potential participants, including through gender-sensitive messages.

15.9 States should implement redistributive reforms through transparent, participatory and accountable approaches and procedures. … All affected parties, including disadvantaged groups, should receive full and clear information on the reforms, including through gender-targeted messages.

Source: FAO, 2012b.

Awareness-raising, gender sensitization and advocacy

People need to be aware of issues and rights so they can exercise both voice and choice effectively (FAO, 2011a: 20). Information must therefore be accessible in local languages (written and spoken) and tailored to audiences at all levels, from national policy-makers to local communities at the grassroots, who may have different opinions, perspectives, experiences and languages.

Awareness raising has an important role in helping to remove barriers to women's access to land information within land sector projects and reforms and throughout all the processes and institutions of land tenure governance.

Awareness raising is most effective when many different stakeholders – including those from government and civil society – collaborate and when entire communities are involved in awareness-raising activities. This requires a carefully thought-out strategy for reaching both women and men and working with diverse stakeholders at different levels.

Effective awareness raising on land and gender issues is particularly important in the implementation of land sector projects and reforms. Examples of good practice are described in Box 5.2. Important factors for success include using carefully targeted interventions, in collaboration with civil society organizations, facilitating women's clear leadership of awareness-raising meetings, and adopting a participatory approach that works with local beneficiary communities.

BOX 5.2:
Good practice in awareness raising

Sources: Renée Giovarelli's presentation on "Gender-equitable awareness raising" and Rebecca Sittie's presentation on "Gender issues in managing and implementing land registration in Ghana" at the May 2011 technical workshop; ILC, 2010e.

In *India*, an awareness-raising project by the local NGO Swadhina in West Bengal's West Midnapore district used a range of awareness-raising activities to involve community members of all levels, with targeted events to boost women's knowledge, confidence and literacy levels. The project's aims were to increase awareness of land rights and promote gender equity among tribal and marginalized rural families. Activities included:

- ten one-day meetings in ten communities, district- and block-level meetings for women and men, and 16 awareness-raising street theatre performances;
- production of communication materials for local and nationwide use, including a short film in the local language on women's empowerment and land rights, posters and cartoon booklets;
- socio-legal leadership training for women and men in the ten communities, and 14 signature camps where 354 illiterate women learned how to sign their names;
- earth festivals in two communities, celebrating women's contribution to local agriculture and involving the participation and mobilization of the whole community.

In the *Lao People's Democratic Republic,* although principles of gender equity are mandated in family, inheritance and property laws, and much of society is customarily matrilineal with no restrictions on women owning land, a World Bank-funded project for the registration of land titles in urban areas performed poorly on gender equity over its first two years: when women inherited land it was their husbands who registered it; very few women participated in project meetings and other activities; the forms used made no provisions for joint ownership and asked for the name of the household head, who is generally the oldest man in the family; and the documentation required to support ownership claims was generally registered in the name of the household head.

To address these issues, the Lao Women's Union held two gender information sessions in each project community – one with both men and women and one with only women. Community training focused on rights under the law and the benefits of involvement in land titling, and Lao Women's Union members started to take part in the project's titling brigades at the local level. This case demonstrates the importance of project monitoring to identify issues and challenges in achieving gender equity, and flexible approaches for addressing them.

In *Ghana*, the Millennium Development Authority's Rural Systematic Titling project, funded by the Millennium Challenge Corporation, included important awareness-raising components on gender and land that helped increase its impact:

- Stakeholder engagement and public education in the communities ensured equal access and opportunities for women and men. The names of all joint owners were registered on land certificates.
- The project adopted a participatory methodology:
  - To reach the grassroots, non-governmental and community-based organizations employed by the project introduced the survey and inventory activities to the communities.

↓

- Land tenure rights were well documented and explained to the communities.
- Communities participated in field mapping exercises.
- Selected community members were trained in alternative dispute resolution techniques for resolving their own land conflicts.
- Follow-up public information sessions clarified outstanding issues.

- Women led the project's awareness-raising and gender sensitization activities in communities, supported by other women and men.

Deeply held attitudes about gender and land may be particularly difficult to change, so specific gender sensitization efforts must be planned and integrated into broader communication and awareness-raising strategies.

Effective gender sensitization should target both women and men, across ages, castes, religions, ethnicities, social status, etc. Gender sensitization empowers women through information and assertiveness skills, but it is also important that men ensure the space for women to negotiate their needs. Achieving this may require tackling issues related to what it means to be a man or a woman in the culture concerned (WOLREC, 2011).

Gender sensitization on land tenure can take various forms: programmes for raising public and community consciousness about women's tenure rights; basic educational measures for informing women and men of their rights; actions that support legal literacy and protect land rights; and programmes that encourage the active participation of women and women's groups in the governance and management of resources (FAO, 2000: 10).

Box 5.3 provides examples of good practice, all of which suggest that effective gender sensitization relies on the innovative use of a wide range of different communication tools and materials.

CHECKLIST 5.1:
Supporting gender-equitable access to land information

Source: FIG, 2001.

- ✓ Explain to women the rights and obligations associated with holding land titles.
- ✓ Discuss the meanings of land administration and management terms with women.
- ✓ Consult women who will be directly affected by land sector project and reform outcomes.
- ✓ Ensure that there is a two-way communication mechanism between women, and surveyors and land sector technicians in the field.
- ✓ Include women surveying professionals in the teams working with local communities.
- ✓ Support women's membership in land institutions at all levels.

REMEMBER!
Barriers to women's access to land information must be removed to create gender-equitable access to land information within land sector projects and reforms.

BOX 5.3:
Gender sensitization in support of gender-equitable land tenure governance

In *Pakistan*, in 2009-2010 a project funded by the International Land Coalition (ILC) monitored the impact on women's access to land of the Government of Sindh's Land Distribution Programme(LDP) and provided legal assistance to women involved in land disputes. An important component of the project was sensitizing the media on land and gender issues, with increased media attention to the LDP contributing to the provincial government's subsequent acknowledgement of flaws in the land distribution process. Sensitization activities included dialogue with the media on the LDP and related issues; media visits to the districts and to land grantees; articles for publication by the media; videos and documentation of case studies and success stories; and the media launch of a study on LDP issues.

Sources: ILC, 2010c; ULA, 2010; World Bank/FAO/IFAD, 2009a: 159–161.

The Land and Equity Movement in *Uganda* (LEMU) is sensitizing communities on land and gender issues and providing information to the public – including on the options for families that are deciding whether or not to title their customary land, or that are facing compulsory purchases. LEMU uses leaflets, radio phone-ins and individual legal advice to disseminate information on women's land rights under customary and statutory law and on the steps communities can take to reduce land conflicts, and to stimulate gender sensitization and debate.

In *Nepal*, the International Fund for Agricultural Development's (IFAD's) Hills Leasehold Forestry and Forage Development Project leases user rights over forest land to groups of five to ten poor households. To ensure that it responds to the needs and interests of rural women, the project hired a three-woman gender team to develop an innovative gender sensitization strategy. Activities included recruiting and training a cadre of women's group promoters to sensitize the women and men beneficiaries of the project; identifying, training and coaching gender focal persons in the project's implementing agencies, which are staffed mainly by men; and producing and distributing two magazines: one for gender focal persons at the district level and the other for group promoters at the grassroots level.

After nine years of implementation, successes included greater participation and leadership roles for women – 25 percent of the participants in leasehold forestry group meetings were women and there were 74 all-women groups and 112 women group leaders; increased self-confidence for the young women involved in the project; and increased land titling to women, with reports of men becoming more accepting of their women's right to hold agricultural land in their own names. Women who had been part of these groups for at least five years reported that 90 percent of all household decisions were now taken jointly.

CHECKLIST 5.2: **Gender sensitization**

Source: FAO, 2002c: 27.

✓	Sensitize both women and men, to change deeply held values and attitudes.
✓	Sensitize national, district and local leaders on gender and land issues.
✓	Promote women's participation in land sector projects and reforms by providing appropriate gender sensitization and training to all project staff, and to members of the local communities involved.
✓	Foster discussion of gender and land issues in a neutral forum, and ensure that effective and culturally appropriate mechanisms are in place for identification of, and discussion, negotiation and conflict resolution on, the issues.
✓	Provide equal opportunities for women and men in gender sensitization, education and training (including study tours) in all land sector projects and reforms.
✓	Facilitate the appointment of qualified women to leadership and decision-making roles in land issues.
✓	Provide special training on gender issues to women, to foster new skills and resources for challenging their traditional roles and participating fully in the processes and institutions of land tenure governance.
✓	Provide special training on gender issues to men, to encourage them to ensure space for women to participate in land tenure governance, by tackling issues related to what it means to be a man or a woman in the culture concerned.
✓	Sensitize the media, to facilitate sustainable change at the community, local government and national levels.

REMEMBER!

Gender sensitization is a necessary but not sufficient part of any successful communication strategy for building sustained support for responsible gender-equitable governance of land tenure. International organizations have an important role in education and sensitization activities for promoting gender integration, gender mainstreaming and gender equity in the processes and institutions of land tenure governance.

Although advocacy work is often directed to policy-makers in government and other decision-making bodies, awareness raising among a wider audience of stakeholders and interest groups can support advocacy efforts by developing public pressure for change and reform.

CHECKLIST 5.3:
Using communication for advocacy

✓	Provide capacity development and financial support to civil society organizations that work locally and/or in partnership with community-based organizations, as they can help bridge the gaps (such as language barriers) between local communities and governments.
✓	Involve the entire community in any communication strategy for gender-equitable governance of land tenure, and stress the benefits of gender equity to all: ‣ Rather than pushing for women's rights in isolation, encourage support for the most marginal and vulnerable people within communities, to avoid direct confrontation on deeply held attitudes about gender relations.
✓	Ensure that all communication products and tools stress gender equity. Consider using different products and tools for reaching men and women, according to the different types of media that they have access to and are interested in.
✓	Pay special attention to addressing rural women's illiteracy issues, and use the whole range of information and communication technologies that women have access to, including cell phones.
✓	Engage women and build their confidence in speaking for themselves in public.
✓	Take into consideration women's workload and time availability for attending meetings.

REMEMBER!
Communication is vital to advocacy work. Well-designed and implemented communication strategies can help advance a cause; poor communication can damage it.

Sources: Sabine Pallas's presentation on "Developing communication and advocacy tools through action-oriented research in Africa" at the May 2011 technical workshop; ILC, 2011.

Legal literacy

Legal literacy for gender-equitable land tenure governance requires that both women and men are aware of what is legally possible and that they have information on how to record land rights, engage in land transactions, and participate in the processes and institutions of land tenure governance. They must also understand the complexity of land issues, the limitations of the law and of the implementation capacity of land administration and management institutions, and the various formal and informal options, mechanisms and legal spheres that may be available for pursuing their land claims. The people charged with implementing land laws and running the institutions of land tenure governance, such as local leaders, judicial authorities and land sector administrators and technicians, must also understand the law and its implications (FAO, 2011a: 18–19; World Bank/FAO/IFAD, 2009a: 144).

Communication and support to improve legal literacy must address the often long-standing customary discrimination that causes some women to give up their statutory land rights, even when they know they have those rights.

Good practices in creating and supporting legal literacy for gender-equitable land tenure governance include the participatory approach used in Colombia and a legal awareness campaign using easily replicable songs, dance and visual tools in very vulnerable and largely illiterate indigenous communities in the eastern Democratic Republic of the Congo, described in Box 5.4.

BOX 5.4:
Supporting legal literacy

Sources: ILC, 2010a; 2010b.

In *Colombia*, a legal literacy project carried out by the NGO Centro de Investigación y Educación Popular in 2010 trained 30 women from rural organizations to increase community awareness of land rights and the mechanisms for claiming them. The participatory approach included:

- participatory research and analysis, leading to production of a document on the challenges women face in obtaining access to land in Colombia;
- three training sessions held in each of five villages, plus a final meeting for learning and experience sharing by all 30 women;
- production of communication materials, including a bulletin and posters on women and access to land, and a radio show broadcast on three regional radio channels;
- drafting and validation of a legal strategy for defending and claiming women's rights to land, approval of a manifesto and dissemination of a toolbox, by the 30 women participants.

These activities had a significant impact on participants' confidence in speaking out and claiming their rights, and on their motivation for collaborating for a common goal.

In the *Democratic Republic of the Congo*, in 2009–2010, the Union pour l'Emancipation de la Femme Autochtone (UEFA) carried out a legal awareness campaign funded by the International Land Coalition (ILC) focusing on women's land rights in pygmy communities in the provinces of North and South Kivu. Activities included:

- capacity-building for ten awareness raisers ("sensibilisateurs") on communication skills and techniques, land and forest law, women's rights and gender and development;
- production of legal awareness materials, including a cartoon booklet and a song in the local language, designed for these largely illiterate communities with a rich but forgotten culture of songs recognizing women's rights and status;
- campaigning activities reaching nearly 9 000 people and 20 local leaders, estimated by UEFA to represent 15 percent of the local pygmy population;
- advocacy with local leaders, using a memorandum on women's rights to land prepared by delegates from the project communities as a basis for future action.

The campaign increased community awareness about land rights among pygmy communities, and local leaders' awareness of the land rights of women pygmies. The use of folk songs, dance and visual tools was particularly valuable in boosting legal literacy among these very vulnerable and marginalized people, and provides a model of good practice.

Communication and awareness-raising strategies and methods

When designing communication and awareness-raising strategies and methods it is important to identify the different ways in which women and men engage with the world and obtain and process information in the country, project and/or reform concerned.

CHECKLIST 5.4:
Designing communication and awareness-raising strategies and methods

- ✓ Where do women and men generally obtain information?
  - For example, schoolteachers are often an important source of information for women, but less so for men.
- ✓ What are the best media for reaching women?
- ✓ What are the social and cultural implications of land rights for each gender? What different impacts might land tenure reforms have for women and men?
- ✓ How and when do different people participate in land tenure governance? What level of knowledge might they already have?
- ✓ What messages need to be given to women and men at and about the different stages of land sector projects or reforms? How should these messages be framed?

REMEMBER!
Communication and awareness-raising efforts risk not having an adequate impact if these questions are not addressed early in the design stage of an information campaign.

The design and implementation of communication actions should address strategic aspects, such as long-term change in values and attitudes; methodological aspects such as choice of methods, tools and messages for different audiences; and institutional aspects, such as gender equity in the organization or institution promoting the action (FAO-Dimitra, 2011b).

Figure 5.1 shows the steps in designing communication and awareness-raising strategies and methods to support gender-equitable land tenure governance.

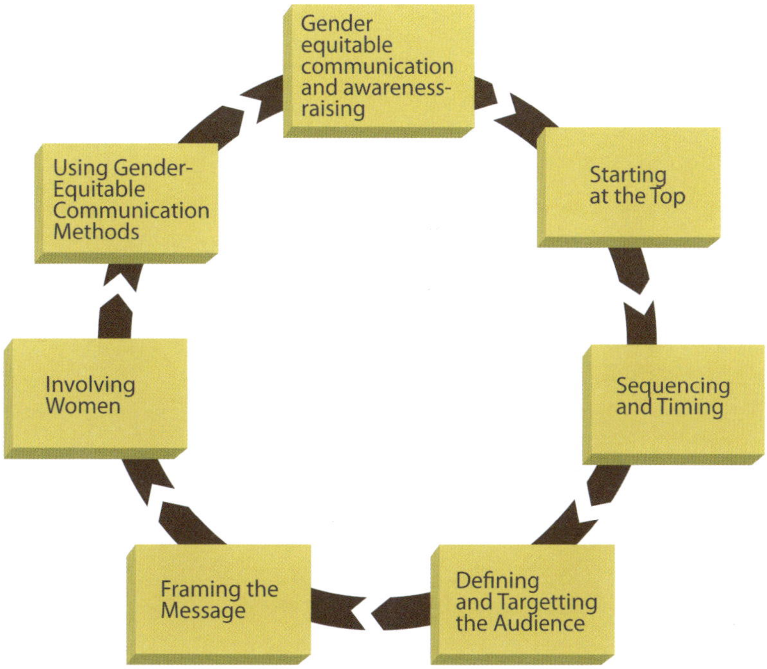

FIGURE 5.1: **Designing a communication and awareness-raising strategy**

Starting at the top

Successful communication and awareness raising depends on having the support and commitment of people at the top levels of decision- and policy-making. Any communication strategy should therefore start by sensitizing and training government officials and land sector administrators and technicians on:

- why gender matters;
- how individual land sector projects and reforms affect women and men – culturally, socially and economically;
- the benefits of gender-equitable land tenure governance, and of involving both women and men in all processes and institutions of land tenure governance, including land administration and management institutions, and project and reform implementation activities;
- best practices for promoting gender equity.

This first step also requires clear understanding of the context and objectives of communication and awareness raising. In terms of the policy-making framework described in *Module 1*, it is helpful to build a mandate for change through a multi-stakeholder alliance, identifying champions and a pioneer group to promote gender equity in land tenure governance and to take the lead in designing and implementing the overall communication and awareness-raising strategy.

Sequencing and timing

Providing the public with information early in the process can help to correct false assumptions and avoid confusion. However, if land institutions are functioning poorly or are corrupt, the information provided may fuel violence and civil discord. This is a particular risk in post-conflict situations, where land and resource issues may have been a contributing factor in the conflict.

This second step in the strategy design process also requires consideration of planning and budgeting for the implementation of communication and awareness-raising activities.

Defining and reaching the target audience

Within the overall communication strategy, different messages need to be developed for different audiences. The same information therefore needs to be presented in different ways, focusing on elements that are of interest and relevance to different groups of people.

It is also critical to assess the target audience's existing level of knowledge and awareness, of both gender, in general, and the specific land tenure issues at stake. This can be achieved by mapping the audience to establish whether it is open to discussing gender issues, the types of communication actions and messages that might be most appropriate, and the likely timeframe. Audience mapping also considers the most useful channels and locations for communication, the activities for communication and awareness raising on gender equity and land issues that other organizations and institutions are carrying out, and the methods and tools that are already being used and can be built on.

Defining the target audience includes asking who/what holds the mandate to build awareness on gender and land tenure governance, and ensuring that these individuals and institutions are included in the audience, and as actors in message delivery.

Framing the message

Messages need to be carefully developed and framed, taking into consideration the target audience, the content of the messages, and how and by whom they will be delivered, as well as the country and project context.

Messages are more powerful when the people delivering them feel that they own them, so it is important to take a participatory approach to developing messages. Tactics are important – presenting change as a positive development that brings benefits for the whole community is likely to be more successful than presenting it as involving loss of power for existing power-holders. The analysis needed for framing messages should therefore include consideration of the problems that gender inequity in land tenure governance currently causes.

Tiny details can also make a dramatic difference to how people interpret messages. For example, a cartoon picturing a man and woman registering their land in which the man is shown signing the form while the woman stands smiling by his side does not convey the message of gender equity as well as one in which the woman is also pictured signing the form.

The community conversation method described in Box 5.5 is a powerful tool for change, enabling the whole community to hear the views of other community members. Community conversations were developed by people working on HIV/AIDS issues and female genital mutilation in Africa, as a method for facilitating major cultural change, and they are just as useful in land issues. However, applying the method in other contexts requires adaptation, as in India, where the idea of community is very different from that found in Africa because of the need to take caste and religion into account, as well as gender. It may therefore not always be possible to work with whole communities and initiate the kind of open community conversation that is possible in some contexts.

BOX 5.5:
Framing messages through community conversations

Source:
Renée Giovarelli's presentation on "Gender-equitable awareness raising" at the May 2011 technical workshop.

A project funded by the Nike Foundation and implemented by the United States NGO Landesa in *West Bengal, India*, adopted a community conversation approach to promote land tenure security for girls and to examine broader issues of girls' inheritance. Landesa organized meetings at which the community and separate groups of boys and girls discussed the issues that make it difficult for girls to control assets: poverty, dowry requirements, early marriage. Because of cultural restrictions on the mixing of different population groups, the project had to hold separate meetings with different caste, religious and other groups before the whole community could meet together.

In conversations with the separate groups, Landesa learned that the project's aim of reducing the dowry payments paid by poor families conflicted with the community's view that small dowries reduced daughters' status. Girls from poorer families often married very young so their families could pay less dowry. Eventually, community members reached consensus that rather than abolishing or reducing dowry payments, a portion of the dowry would be viewed as an asset for their daughters. The community conversations also brought to light a demand from both mothers and daughters for land-based vocational training to develop skills with which they could earn an income. As a result of this approach, the project was able to adapt its support to adolescent girls from homeless and landless families to reflect their interests and needs within the community.

CHECKLIST 5.5:
Message development

Sources: Sabine Pallas's presentation on "Developing communication and advocacy tools through action-oriented research in Africa" at the May 2011 technical workshop; ILC, 2011.

- ✓ Identify the best media for reaching women.
- ✓ Know the political environment – draw a power map to identify key stakeholders.
- ✓ Know the audience – carry out audience mapping.
- ✓ Keep messages simple and brief.
- ✓ Be persuasive, not aggressive.
- ✓ Use real-life stories and quotations.
- ✓ Use clear facts and numbers creatively.

REMEMBER!
Different audiences have different communication needs. The language will need to be adjusted and the messages tailored accordingly for the best understanding and uptake.

Involving women

The location and timing of meetings are crucial for involving women. For example, women often have a lot of work to do in the home, so it might be appropriate to hold meetings for small groups of neighbouring women in one of the women's houses. Inviting women to bring their families' food to meetings so that the cooking can be done collectively during the meeting can also help boost women's participation, as illustrated in the example from Uganda in Box 5.6.

It may be best to hold meetings for women in the evening, after they have finished their work for the day. However, evening meetings may be inappropriate where women live far apart, as their safety and security might be endangered by walking home in the dark. It is particularly important to keep messages simple and clear in evening meetings, so the meeting can end before everyone gets tired.

It is also important to hold meetings in locations where women can comfortably attend. Project meetings can be combined with existing meetings involving women, or be held at places where women often congregate, such as government water points, tea houses or – as in Malawi (Box 5.6) - literacy centres.

BOX 5.6:
Supporting women's participation in meetings

Sources: Rachael Knight's presentation on "Gender-equitable participation in land governance: community land titling in Liberia, Mozambique and Uganda", and discussions of the communication working group at the May 2011 technical workshop; World Bank/FAO IFAD, 2009b: 441.

An agroforestry project in Gudibanda Taluk, Karnataka, *India* – implemented by the NGO Women For Sustainable Development – supports local women and men farmers in planting orchards for harvesting and carbon sequestration. The project facilitates women's participation in decision-making processes by taking into account women's time and cultural constraints when establishing public fora.

During implementation of the International Development Law Organization's Community Land Titling project in Okere, *Uganda*, women reported that they had to miss many meetings as they were working and did not have time. So a community support person asked women to send their food to the meeting place, then cooked all the food, ensuring that the women attended the meeting.

An award-winning ActionAid project in *Malawi* incorporated discussion of gender and land issues into weekly literacy classes for women.

Using gender-equitable communication methods

Some communication methods and media are more effective than others in reaching different audiences, depending on the availability of resources, time and control over access and use.

For example, radio is an important tool for awareness raising, agricultural extension and information dissemination in rural areas and isolated communities. Radios can also be used in participatory approaches, as in FAO's Dimitra project described in Box 5.7. The Dimitra approach is a good practice for use in rural areas, to encourage debate and discussion. However, in some cases, men control use of and access to household radios.

BOX 5.7:
Rural radios as a communication tool for change – the Dimitra project

Dimitra is an FAO project and network on gender equality, information, communication and food security in rural areas. With partners in *the Niger* and the *Democratic Republic of the Congo* (DRC), in 2006, Dimitra developed community listeners' clubs, which now have about 8 000 members in the two countries.

Separate and mixed groups of women and men villagers meet regularly to discuss their concerns and challenges and take collective action for addressing them. Rural radio stations relay and exchange information, providing broader access to knowledge and raising awareness on the themes identified by club members: agricultural practices, food security, access to land, sanitation, etc. Club members use solar-powered and wind-up radios and (sometimes) cellular phones to communicate with each other and with other clubs. Through dialogue and action, rural people are better informed, share knowledge, gain self-confidence and speak out. This process also encourages them to participate actively in community life.

Some of these results have been particularly helpful in promoting gender equity. For example, in the Niger, women in the village of Banizoumbou negotiated with local men landowners to obtain a 99-year lease on 2.75 ha of good-quality land for growing vegetables. In the DRC, from discussions among listeners' clubs in Bugobe (Kabare),it emerged that men were taking over fertile wetlands to plant sugarcane. This had significant impacts on households' food situation, as it deprived women of land for growing other crops. The listeners' club alerted administrative and traditional authorities, which imposed extremely high taxes on the sugar-cane growers. As a result, about 70 percent of the land was returned to fruit and vegetable cultivation.

These examples show how better access to information and communication can stimulate questioning of practices that are not gender-equitable and discussion of important community development issues about agriculture, food security, women's rights, etc., leading to social mobilization and empowerment of community members, particularly women. One man described how his involvement in a listeners' club in Kalehe in South Kivu, DRC, was the first time that he had participated "in a debate on gender awareness. I had always thought this was a topic that favoured women and went against men!"

Sources: FAO-Dimitra, 2008; 2010a; 2010b; 2011a; 2011b; 2011c; Eliane Najros, personal communication.

Rural women often have very limited access to print media and printed information, so oral and visual communication tools should be particularly stressed. Messages and information may also need to be translated into local languages, and should avoid technical or legal jargon. Community theatre, comic books, songs, posters and videos are additional tools, as described in Box 5.8 about a project in Malawi.

BOX 5.8:
Gender-equitable communication methods and media

Action with traditional leaders, who can be very effective in bringing about social change when they are aware of and recognize the legitimacy of gender issues and problems.

Video documentation, capturing songs, poems, dramas and different voices, although people sometimes do not want to face the camera and have their identities revealed.

↓

Source: WOLREC, 2011.

Theatre for development, focusing on land issues where gender inequalities arise, and disseminating research findings within the community, giving audiences an opportunity to discuss their realities and find solutions to challenges regarding land.
Peer-to-peer exchange visits between patrilineal and matrilineal communities, to compare their land tenure systems – this is a resource-intensive communication method.
Community visioning exercises, in which community women and men draw maps of what women's ownership of and control over land would look like.

Changing values and attitudes

Improving gender equity in the governance of land tenure is a long-term project requiring changes in fundamental and often deeply held values and attitudes. To have the greatest impact over the longer term, it is important to identify entry points for communication and awareness raising on gender and land issues. Effective strategies include working with children, whose values and attitudes will become those of future societies; the community conversation approach described in Box 5.5; and the family-oriented approach described in Box 5.9.

BOX 5.9: **Changing attitudes through a family-oriented approach**

Source: World Bank/FAO/IFAD, 2009a: 162–165.

The World Bank's Land Access Pilot Project and the expanded Proyecto Acceso a la Tierra were implemented in *Honduras* between 2001 and 2009 to promote land acquisition by poor people, increased awareness of joint property rights, legal guarantee of rights regardless of marital status, equal participation by household members in rural small businesses, women's management of new businesses, and gender-equitable training. It provided landless and land-poor rural families with technical and legal assistance and land purchase loans, taking into account the activities of wives who support their husbands' business ventures. The project was designed with a participatory monitoring and evaluation system, and incorporated a gender strategy from 2003. All technical staff were trained and sensitized on gender equity.

Project staff reported that implementing a gender strategy while addressing cultural differences was challenging but produced successful, gender-equitable project outcomes. A major long-term lesson is that providing family-oriented information and awareness encourages men to change their attitudes towards family obligations, value the contributions that their wives or partners make to production activities, and recognize their wives' or partners' ownership rights over any assets they may acquire.

Communication and awareness raising for gender-equitable land tenure governance is an ongoing and iterative process. Audience maps need to be updated and progress monitored against measurable objectives and indicators. Donors and funders need to be sensitized on the long-term nature of the process of changing values and attitudes, so that resources for implementing responsible gender-equitable governance of land tenure will be forthcoming.

Notes

[18] Patricia Costas, Catherine Gatundu, Renée Giovarelli, Paolo Groppo, Rachael Knight, Sabine Pallas and Rebecca Sittie provided substantial inputs to this module during the May 2011 technical workshop at FAO Headquarters, including through discussions of the communication working group.

Summary of key messages on getting the message across

Communication and awareness raising are key to mainstreaming gender issues and achieving responsible governance of land tenure.

Attitudes towards gender-equitable governance of land tenure may be difficult to change because they are deeply held.

Changing values and attitudes is therefore a long-term project requiring much time and effort, and sensitization of donors and funders to ensure the availability of resources for supporting communication and awareness-raising strategies.

Gender sensitization is needed for women and men, local communities and the media, and participants at all levels in the processes and institutions of land tenure governance.

Effective awareness-raising strategies are pragmatic, flexible, adaptive and collaborative, involving different stakeholders in a participatory way.

Successful communication and awareness raising to support gender-equitable governance of land tenure must be carefully designed, starting from the top, considering sequencing and timing, defining and targeting the audience, developing and framing the message, and involving women and men at all stages.

Gender-equitable communication methods and tools include community conversations, rural radio, and participatory and visual tools.

Advocacy and legal literacy campaigns are also important communication activities.

Communication and awareness raising to support gender-equitable land tenure governance is an ongoing and iterative process.

Suggested monitoring indicators for getting the message across

- number of gender sensitization and awareness-raising initiatives focusing on land
- number of gender-sensitive advocacy and legal literacy training sessions focusing on land
- % of women and men receiving advocacy and legal literacy training focusing on land
- number of communication tools and media used to get the message across on gender-equitable land tenure governance
- numbers of women and men reached by the messages

Aksornkool, N., Joerger, C. & Taylor, E., eds. 2004. *Gender sensitivity: a training manual for sensitizing education managers, curriculum and material developers and media professionals to gender concerns.* Paris, UNESCO. http://unesdoc.unesco.org/images/0013/001376/137604eo.pdf

FAO. 2000. Integrating gender in land and natural resource tenure programmes: Policy guidelines, S. Lastarria-Cornhiel. Rome. (unpublished)

FAO. 2002a. *Gender and access to land. Land Tenure Studies No. 4.* Rome. ftp://ftp.fao.org/docrep/fao/005/y4308e/y4308e00.pdf

FAO. 2012b. Voluntary Guidelines on the Responsible Governance of Tenure of Land, Fisheries and Forests in the Context of National Food Security. Rome. http://www.fao.org/docrep/016/i2801e/i2801e.pdf

FAO-Dimitra.2008. *Accès à la terre en milieu rural en Afrique: stratégies de lutte contre les inégalités de genre*, Atelier FAO-Dimitra, 26–28 septembre 2008. Brussels.

FAO-Dimitra. 2010a. Projet club d'écoute: Rapport d'activités, 2ème étape: Janvier–Août 2010, H.A. Adamou. Brussels.

FAO-Dimitra. 2010b. Rapport: mission d'accompagnement du Project Clubs d'écoute pour l'automatisation et le leadership des femmes rurales et des jeunes des centre d'alphabétisation du 19 au 22 mai 2010. Brussels.

FAO-Dimitra. 2011a. Communiquer le genre pour le développement rural. Rome.

FAO-Dimitra. 2011b. Community listeners' clubs. Stepping stones for action in rural areas. Rome. www.fao.org/docrep/014/am604e/am604e.pdf

FAO-Dimitra. 2011c. Niger: Land for a century for Banizoumbou's women. *Newsletter*, 21:3–4. Brussels. www.fao.org/docrep/015/an604e/an604e00.pdf

FAO/GTZ. 2006. Framework on effective rural communication for development. Rome and Eschborn, Germany. www.fao.org/nr/com/gtzworkshop/a0892e00.pdf

FIG. 2001. Women's access to land – FIG guidelines. Principles for equitable gender inclusion in land administration: Background report and guidelines. FIG Publication No. 24.Copenhagen, International Federation of Surveyors (FIG).

ILC. 2010a. Women's Access to Land – Project Brief 1: Strengthening the awareness of authorities and communities of the tenure rights of indigenous pygmy women in the provinces of South and North Kivu. Rome. (unpublished)

ILC. 2010b. Women's Access to Land – Project Brief 2: Legal empowerment of women to claim land and territorial rights in the Cauca Valley, Colombia. Rome. (unpublished)

ILC. 2010c. Women's Access to Land – Project Brief 3: Promoting women's property rights and ownership in Pakistan. Rome. (unpublished)

ILC. 2010e. Women's Access to Land – Project Brief 5: Bringing land-rights to the grassroots women: creating awareness on socio-legal rights in West Midnapore district of West Bengal in India. Rome. (unpublished)

ILC. 2011. Securing Women's Access to Land Project: Advocacy Toolbox. www.landcoalition.org/sites/default/files/advocacy_toolbox_eng.pdf

ULA. 2010. *The Uganda Land Alliance quarterly newsletter, Vol. 4.* Kampala.

WOLREC. 2011. *Women's access to land and household bargaining power: a comparative action research project in patrilineal and matrilineal societies in Malawi.* Research Report No. 9. Rome, ILC, Women's Legal Resource Centre. www.landcoalition.org/sites/default/files/publication/959/wlr_9_malawi.pdf

World Bank/FAO/IFAD. 2009a. Module 4: Gender issues in land policy and administration. *In Gender in agriculture sourcebook.* Washington, DC and Rome.

World Bank/FAO/IFAD. 2009b. Module 10: Gender and natural resources management. *In Gender in agriculture sourcebook.* Washington, DC and Rome.

Resources and references

Further support for gender-equitable land tenure governance

Introduction

The pool of available resources that support the pursuit of responsible gender-equitable governance of land tenure is constantly expanding. This guide is therefore not intended to be fully comprehensive.

This section sets out the full references of all the documentary sources cited in this guide, and provides links to additional resources on the Internet. Users of the guide are also referred to the Convention on the Elimination of All Forms of Discrimination against Women (CEDAW) and other international and regional instruments for additional guidance in the pursuit of responsible land tenure governance that is gender-equitable. Information about some of these instruments is at the end of this section.

Presentations at the May 2011 technical workshop

Adoko, Judy, "Constructive ways of working with customary institutions in support of women's Land rights".

Alnaggar, Dalal, "Gender-equitable capacity in technology and land survey in the Arab region".

Costas, Patricia, "Gender-equitable communication methods and sensitization strategies to support land governance in Bolivia".

Dalrymple, Kate, "Gender and technology issues in land administration projects in Southeast Asia and the Pacific".

Edmead, Nigel, "Gender issues in designing and setting-up land information systems and databases: experiences from Ghana, Zambia and Uganda".

Fletcher, Debra, "Land administration – improving gender equity in institutions and activities: Aboriginal participation in land management in Australia with emphasis upon Western Australia".

Gaikwad, Seema, "Asian experiences of civil society engagement with land policy processes".

Gatundu, Catherine, "Africa experiences of civil society involvement in land policy process".

Giovarelli, Renée, "Gender-equitable awareness raising".

Knight, Rachael, "Gender-equitable participation in land governance: community land titling in Liberia, Mozambique and Uganda".

Pallas, Sabine, "Developing communication and advocacy tools through action-oriented research in Africa".

Sittie, Rebecca, "Gender issues in managing and implementing land registration in Ghana".

Stair, Elizabeth, "Land and gender – the Jamaican context".

Stowers-Fiu, Josephine, "Capacity building and gender equity in land administration institutions in Samoa".

References and further reading

Adams, M., Berkoff, J. & Daley, E. 2006. *Land-water interactions: opportunities and threats to water entitlements of the poor in Africa for productive use.* Human Development Report Occasional Paper No. 2006/20. Washington, DC, UNDP. http://hdr.undp.org/en/reports/global/hdr2006/papers/adams%20martin%20et%20al.pdf

ADB. 2006. *Republic of Uzbekistan: land improvement project.* Report and Recommendation of the President to the Board of Directors, Proposed Loans and Technical Assistance Grant. Manila, Asian Development Bank. www2.adb.org/documents/rrps/uzb/37536-01-uzb-rrp.pdf

Agarwal, B. 2009. Gender and forest conservation: The impact of women's participation in community forest governance. *Ecological Economics*, 68(11): 2785–2799.

Ahikire, J. 2010. *Half full or half empty glass. Decentralisation and women's land rights in Uganda.* Paper presented at the Women's Land Rights Conference, Speke Resort Munyonyo, Uganda, 4–6 October 2010.

Aksornkool, N., Joerger, C. & Taylor, E., eds. 2004. *Gender sensitivity: a training manual for sensitizing education managers, curriculum and material developers and media professionals to gender concerns.* Paris, UNESCO. http://unesdoc.unesco.org/images/0013/001376/137604eo.pdf

Alden Wily, E. 2003. *Governance and land relations: A review of decentralisation of land administration and management in Africa.* London, IIED. http://pubs.iied.org/pdfs/9304IIED.pdf

Ambaye, D. W. 2009. Land valuation for expropriation in Ethiopia: Valuation methods and adequacy of compensation. Paper presented at the 7th FIG Regional Conference on Spatial Data Serving People: Land Governance and the Environment – Building the Capacity, Hanoi, 19–22 October 2009.

Aristazabal, N. & Gomez, A. 2004. Improving security without titles in Bogotá. *Habitat International*, 28(2): 245–258.

Augustinus, C. & Benschop, M. 2003: Security of tenure – best practices. Nairobi, UN-HABITAT. http://www.unhabitat.org/downloads/docs/10784_1_594339.pdf

Barnett, K. & Grown, C. 2004. Gender impacts of government revenue collection: The case of taxation. *Economic Paper No. 62.* London, Commonwealth Secretariat.

Binswanger-Mkhize, H.P., Bourguignon, C. & van den Brink, R. 2009. *Agricultural land redistribution: toward greater consensus.* Washington, DC, World Bank.

Bernstein, A. 2005. *Land reform in South Africa – a 21st century perspective. Research Report No. 14.* Johannesburg, Centre for Development and Enterprise.

Bicchieri, M. 2010. Gender and Land Project – Mozambique. Paper presented at FAO Headquarters, Rome, 13May 2010.

Binswanger-Mkhize, H.P., Bourguignon, C. & van den Brink, R. 2009. Agricultural land redistribution: toward greater consensus. Washington, DC, World Bank.

Carfield, M. 2011. Land justice in Uganda: Preserving peace, promoting integration. *In* E. Harper, ed. *Working with customary justice systems: post-conflict and fragile states, pp. 127-–143.* Rome, International Development Law Organization.

Daley, E. 2011. *Gendered impacts of commercial pressures on land.* Rome, ILC. www.landcoalition.org/sites/default/files/publication/902/mokoro_gender_web_11.03.11. pdf

Daley, E. Dore-Weeks, R. & Umuhoza, C. 2010. Ahead of the game: Land tenure reform in Rwanda and the process of securing women's land rights. *Journal of Eastern African Studies*, 4(1): 131–152.

Daley, E. & Englert, B. 2010. Securing land rights for women. *Journal of Eastern African Studies*, 4(1): 91–113.

Djusaeva, S. 2012. *Promoting rural women's rights to land: UN Women programme experience in Kyrgyzstan and Tajikistan.* Paper prepared for the Expert Group Meeting on Good Practices in Realizing Women's Rights to Productive Resources, with a Focus on Land, organized by UNWomen in collaboration with the United Nations Office of the High Commissioner for Human Rights, Geneva, 25–27 June 2012.

Edmead, N. 2011. *Open title: a model for low-cost, customizable land administration technology.*
Paper presented at the Annual World Bank Conference on Land and Poverty, Washington, DC,
18–20 April 2011.

Englert, B. & Daley, E. 2008. Introduction. *In E. Daley and B. Englert. Women's land rights and privatization in Eastern Africa.* Woodbridge, UK, Boydell and Brewer.

FAO. 2000. Integrating gender in land and natural resource tenure programmes: Policy guidelines, S. Lastarria-Cornhiel. Rome. (unpublished)

FAO. 2001. *Good practice guidelines for agricultural leasing arrangements. Land Tenure Studies No. 2.* Rome.

FAO. 2002a. *Gender and access to land. Land Tenure Studies No. 4.* Rome. ftp://ftp.fao.org/docrep/fao/005/y4308e/y4308e00.pdf

FAO. 2002b. *Land tenure and rural development. Land Tenure Studies 3.* Rome. ftp://ftp.fao.org/docrep/fao/005/y4307e/y4307e00.pdf

FAO. 2002c. *Law-making in an African context: The 1997 Mozambican Land Law,* C. Tanner. FAO Legal Papers Online No. 26, Rome.www.fao.org/fileadmin/user_upload/legal/docs/lpo26.pdf

FAO. 2003a. *SEAGA Socio-Economic and Gender Analysis Programme macro level handbook.* Rome. www.fao.org/docrep/012/ak229e/ak229e00.pdf

FAO. 2003b. *The design of land consolidation pilot projects in Central and Eastern Europe. Land Tenure Studies No. 6.* Rome. ftp://ftp.fao.org/docrep/fao/006/y4954e/y4954e00.pdf

FAO. 2004. *Leasing agricultural land. Land Tenure Notes No. 1.* Rome. ftp://ftp.fao.org/docrep/fao/007/y5513e/y5513e00.pdf

FAO. 2006a.*Gender and law: Women's rights in agriculture, L. Cotula.* FAO Legislative Study No. 76, Rev.1.Rome. ftp://ftp.fao.org/docrep/fao/005/y4311e/y4311e00.pdf

FAO. 2006b. *Land tenure alternative conflict management. Land Tenure Manuals No. 2.* Rome. ftp://ftp.fao.org/docrep/fao/009/a0557e/a0557e00.pdf

FAO. 2007. *Good governance in land tenure and administration. Land Tenure Studies No. 9.* Rome. ftp://ftp.fao.org/docrep/fao/010/a1179e/a1179e00.pdf

FAO. 2008a. *Compulsory acquisition of land and compensation. Land Tenure Studies No. 10.* Rome. ftp://ftp.fao.org/docrep/fao/011/i0506e/i0506e00.pdf

FAO. 2008b. *Good governance and natural resources tenure in South East Asia region,* O. Nabangch ang and E. Srisawalak. *Land Tenure Working Paper No. 4.* Rome.

FAO. 2009a. *Participatory land delimitation, an innovative development model based upon securingrights acquired through customary and other forms of occupation,* C. Tanner, P. De Wit and S.Norfolk. *Land Tenure Working Paper* No. 13.Rome. ftp://ftp.fao.org/docrep/fao/012/ak546e/ak546e00.pdf

FAO. 2009b. *Towards voluntary guidelines on re-sponsible governance of tenure of land and other natural resources – discussion paper.* Land Tenure Working Paper *No. 10.* Rome. ftp://ftp.fao.org/docrep/fao/012/i0955e/i0955e00.pdf

FAO. 2009c. "We all inherited this earth", inheritance rights of widows and children in the context of AIDS: legal overview. Rome. (unpublished report)

FAO. 2010. Land tenure, land policy, and gender in rural areas, by S. Lastarria-Cornhiel and A. Manji. Rome. (unpublished)

FAO. 2011a. *Governing land for women and men – gender and voluntary guidelines on responsible governance of tenure of land and other natural resources,* E. Daley and C.M. Park. *Land Tenure Working Paper No. 19.* Rome. http://www.fao.org/docrep/014/ma811e/ma811e00.pdf

FAO. 2011b. *State of Food and Agriculture 2010–11: Women in agriculture, closing the gender gap for development.* Rome. www.fao.org/docrep/013/i2050e/i2050e.pdf

FAO. 2012a. Improving Gender Equality in Territorial Issues (IGETI).Land and Water Division Working Paper No. 3. Rome. www.fao.org/docrep/016/me282e/me282e.pdf

FAO. 2012b. *Voluntary Guidelines on the Responsible Governance of Tenure of Land, Fisheries and Forests in the Context of National Food Security.* Rome. http://www.fao.org/docrep/016/i2801e/i2801e.pdf

FAO-Dimitra. 2008. *Accès à la terre en milieu rural en Afrique: stratégies de lutte contre les inégalités de genre,* Atelier FAO-Dimitra, 26–28 septembre 2008. Brussels.

FAO-Dimitra. 2010a. Projet club d'écoute: Rapport d'activités, 2ème étape: Janvier–Août 2010, H.A. Adamou. Brussels.

FAO-Dimitra. 2010b. Rapport: mission d'accompagnement du Projet Clubs d'écoute pour l'autonomisation et le leadership des femmes rurales et des jeunes des centres d'alphabétisation du 19 au 22 mai 2010. Brussels.

FAO-Dimitra. 2011a. Communiquer le genre pour le développement rural. Rome.

FAO-Dimitra. 2011b. Community listeners' clubs. Stepping stones for action in rural areas. Rome. www.fao.org/docrep/014/am604e/am604e.pdf

FAO-Dimitra. 2011c. Niger: Land for a century for Banizoumbou's women. *Newsletter*, 21:3–4. Brussels. www.fao.org/docrep/015/an604e/an604e00.pdf

FAO/GTZ. 2006. Framework on effective rural communication for development. Rome and Eschborn, Germany. www.fao.org/nr/com/gtzworkshop/a0892e00.pdf

FAO, Norwegian Refugee Council, UNHCR, UN-HABITAT, OHCHR & OCHA. 2007. *Handbook on housing and property restitution for refugees and displaced persons – implementing the "Pinheiro Principles".* www.ohchr.org/Documents/Publications/pinheiro_principles.pdf

Farfan, F. 2002. *In* G. Payne, ed. *Land rights and innovation: improving tenure security for the urban poor.* London, ITDG Publishing.

FIG. 2001. Women's access to land – FIG guidelines. Principles for equitable gender inclusion in land administration: Background report and guidelines. FIG Publication No. 24.Copenhagen, International Federation of Surveyors (FIG).

Fitzpatrick, D. 2008. *Women's rights to land and housing in tsunami-affected Aceh, Indonesia. AcehWorking Paper No. 3.* Singapore, Asia Research Institute and Oxfam International. www.ari.nus.edu.sg/docs/downloads/aceh-wp/acehwps08_003.pdf

Flintan, F. 2010a. *Learning by doing: Working towards participatory rangeland management (PRM) in pastoral areas through ELMT/ELSE.* Washington, DC, USAID.

Flintan, F. 2010b. Sitting at the table: Securing benefits for pastoral women from land tenure reforms in Ethiopia. *Journal of Eastern African Studies*, 4(1): 153–178.

Flintan, F. & Cullis, A., compilers. 2010. *Introductory guidelines to participatory rangeland management in pastoral areas.* Ethiopia, Save the Children USA, FAO and ECHO. www.cop-ppld.net/fileadmin/user_upload/cop-ppld/items/introductory%20guidelines%20prm.pdf

Ghana Land Administration Project. no date. *An overview of LAP's gender initiative.*www.ghanalap.gov.gh/index1.php?linkid=154&sublinkid=161

Giovarelli, R. & Duncan, J. 1999. Women and land in Eastern Europe and Central Asia. Washington, DC, USAID. http://pdf.usaid.gov/pdf_docs/pnacl400.pdf

Giovarelli, R. & Lastarria-Cornhiel, S. 2006. *Study on women and property rights: project best practices.* Washington, DC, USAID. http://pdf.usaid.gov/pdf_docs/pnadj420.pdf

Government of Samoa. 2008. *Legislative drafting handbook.* Legislative drafting requirements approved by the Attorney General for use in Samoa from July 2008. Apia, Legislative Division of the Office of the Attorney General.

GROOTS Kenya. 2011. Complementing the state: the contribution of the "watchdog groups" in protecting women's land rights in Gatundu District, Kenya. Rome, ILC.

Grover, R. & Flores-Bórquez, M. 2004. *Restitution and land markets.* Paper presented at the FIG Working Week, Athens, 22–27 May 2004.

Guyer, J. 1987. Women and the state in Africa: Marriage law, inheritance and resettlement. *Working Papers in African Studies No. 129.* Boston, Massachusetts, USA, African Studies Centre, Boston University.

Harrington, A. & Choprah, T. 2010. *Arguing traditions. Denying Kenya's women access to land rights.* Justice for the Poor Research Report No. 2/2010.Washington, DC, World Bank.

Hilhorst, T. 2010. Decentralization, land tenure reforms and local institutional actors: building partnerships for equitable and sustainable land governance in Africa. *Land Tenure Journal*, 1(10): 35–59.

ILC. 2009. *Advancing the monitoring of land governance for ensuring impact on poverty reduction.* Paper presented atLand Governance in Support of the MDGs: Responding to New Challenges, World Bank, Washington, DC, 9–10 March 2009.

ILC. 2010a. Women's Access to Land – Project Brief 1: Strengthening the awareness of authorities and communities of the tenure rights of indigenous pygmy women in the provinces of South and North Kivu. Rome. (unpublished)

ILC. 2010b. Women's Access to Land – Project Brief 2: Legal empowerment of women to claim land and territorial rights in the Cauca Valley, Colombia. Rome. (unpublished)

ILC. 2010c. Women's Access to Land – Project Brief 3: Promoting women's property rights and ownership in Pakistan. Rome. (unpublished)

ILC. 2010d. Women's Access to Land – Project Brief 4: Sustainable empowerment of tribal communities through land rights movement with women partnership for securing dignity and equity. Rome. (unpublished) International Justice Mission. Our Work: What We Do.http://www.ijm.org/ourwork/whatwedo

ILC. 2011. Securing Women's Access to Land Project: Advocacy Toolbox. www.landcoalition.org/sites/default/files/advocacy_toolbox_eng.pdf

ILC. 2012. *Women's legal empowerment: lessons learned from community-based activities,* S. Pallas and L. Miggiano. ILC Briefing Note.www.landcoalition.org/publications/womens-legalempowerment-lessons-learned-community-based-activities

India Ministry of Rural Development. 2009. *Committee on State Agrarian Relations and Unfinished Task of Land Reforms Vol. 1 – Draft Report*. New Delhi, Government of India.

International Justice Mission. Our Work: What We Do. http://www.ijm.org/ourwork/whatwedo

IPC. 2010. *Civil Society Consultation on FAO's Voluntary Guidelines on Responsible Land and Natural Resources Tenure, Kuala Lumpur, Malaysia, 24–26 March 2010, Final Report*. Rome, International Planning Committee for Food Sovereignty.

Kapur, A. 2011. Two faces of change: The need for a bi-directional approach to improve women's land rights in plural legal systems. *In* E. Harper, ed. *Working with customary justice systems: Post-conflict and fragile states,* pp. 73–92. Rome, International Development Law Organization.

Knight, R. 2011. The community land titling initiative: An investigation into the protection of customary land claims. *In* E. Harper, ed. *Working with customary justice systems: post-conflict and fragile states*, pp. 145–170. Rome, International Development Law Organization.

Khudayberdiyeva, M. 2009. *Uzbekistan Land Improvement Project.*Tashkent, ADB Uzbekistan.

Land Watch Asia. 2010. *Nepal: renewed push for land reform as a new chapter in history unfolds. Campaign Update 2010–2011*. www.angoc.org/portal/wp-content/uploads/2012/06/05/land-watch-asia-campaign-updates/nepal_final.pdf

Lemmen, C. 2010. *The social tenure domain model: A pro-poor land tool.* Copenhagen, FIG. www.fig.net/pub/figpub/pub52/figpub52.pdf

Lunnay, C. 2006. *Surveying and land administration – sustainable tertiary education for developing countries.* Paper presented at the XXIII FIG Congress, Shaping the Change, Munich, Germany, 8–13 October 2006.

Mackenzie, F. 1989. Land and territory: the interface between two systems of land tenure, Murang'a District, Kenya. Africa, 59(1): 91–109.

Mackenzie, F. 1990. Gender and land rights in Murang'a District, Kenya. *Journal of Peasant Studies*, 17(4): 609–643.

Mackenzie, F. 1993. A piece of land never shrinks: Reconceptualizing land tenure in a smallholding district, Kenya. *In* T. Bassett and D. Crummey, eds. *Land in African agrarian systems.* Madison, Wisconsin, USA, University of Wisconsin Press.

MacKinnon, C. 1987. *Feminism unmodified: Discourses on life and law.* Cambridge, Massachusetts, USA, Harvard University Press.

Magezi, J. 2010. *The role of case law in upholding women's land Rights.* Paper presented at the Women's Land Rights Conference, Speke Resort Munyonyo, Uganda, 4–6 October 2010.

McAuslan, P. 2010. Personal reflections on drafting laws to improve women's access to land: Is there a magic wand? Journal of Eastern African *Studies*, 4 (1): 114–130.

Meertens, D. 2012. *Colombia's law on victims and restitution: a challenge for gender-sensitive transitional justice.* Paper presented at the Expert Group Meeting on Good Practices in Realizing Women's Rights to Productive Resources, with a Focus on Land, UN Women in collaboration with the United Nations Office of the High Commissioner for Human Rights, Geneva, 25–27 June 2012.

Mitchell, D. 2010. Land tenure and disaster risk management. *Land Tenure Journal,* 1(10): 121–141.

Moore, H. 1994. The divisions within: sex, gender and sexual difference. *In A passion for difference – essays in anthropology and gender*. Oxford, UK, Polity Press.

Namutebi, M. 2010. *The Office of the Administrator General and women's land rights.* Paper presented at the Women's Land Rights Conference, Speke Resort Munyonyo, Uganda, 4–6 October 2010.

Osskó, A. 2002. Land restitution and compensation procedures in Central Eastern Europe. Paper Presented at the FIG Commission 7 Annual Meeting, Pretoria, 4–9 November 2002.

Rabley, P. 2010. Alleviating poverty in the developing world – leveraging property rights with geospatial technology. *The American Surveyor, January 2010.* www.amerisurv.com/pdf/theamericansurveyor_rabley-alleviatingpoverty_January2010.pdf

RWN. 2011. *Experiences of women in asserting their land rights: the case of Bugesera District, Rwanda.* Rome, Rwanda Women's Network, ILC.www.landcoalition.org/sites/default/files/publication/956/wlr_5_rwn.pdf

Sayers, R. 2006. Principles of awareness raising for information literacy, a case study. Bangkok, UNESCO. http://unesdoc.unesco.org/images/0014/001476/147637e.pdf

Seeley, J., Batra, M. & Sarin, M. 2000. *Women's participation in watershed development in India.* Gatekeeper Series No. 92. London, IIED. http://pubs.iied.org/pdfs/6347IIED.pdf

South Africa Department of Land Affairs. 1997. White Paper on South African Land Policy. Pretoria.

Ssonko, M. N. 2010. *The application of succession law in the realization of women's land rights in Uganda.* Paper presented at the Women's Land Rights Conference, Speke Resort Munyonyo, Uganda, 4–6 October 2010.

Teklu, A. 2005. Research report 4 – land registration and women's land rights in Amhara Region, Ethiopia. London, IIED.

Tsikata, D. 2003. Securing women's interests within land tenure reforms: recent debates in Tanzania. In S. Razavi, ed. *Agrarian change: Gender and land rights.* Oxford, UK, Blackwell.

ULA. 2010. *The Uganda Land Alliance quarterly newsletter, Vol. 4.* Kampala.

UNCHS. 1999. Women's rights to land, housing and property in post-conflict situations and during reconstruction: A global overview. Nairobi, United Nation Centre for Human Settlements.www.unhabitat.org/downloads/docs/1504_59744_Land.pdf2.pdf

Undeland, A. 2012. *Lessons learned and good practices in realizing women's rights to productive resources with a focus on land: Kyrgyzstan and Tajikistan.* Paper prepared for the Expert Group Meeting on Good Practices in Realizing Women's Rights to Productive Resources, with a Focus on Land, organized by UN Women in collaboration with the United Nations Office of the High Commissioner for Human Rights, Geneva, 25–27 June 2012.

UN-HABITAT. 2007. Gendering land tool development. Nairobi, GLTN Secretariat.

UN-HABITAT. 2007b. Policy makers guide to women's land, property and housing rights across the world. Nairobi.

UN-HABITAT. 2008. Gendering land tools: Achieving secure tenure for women and men. Nairobi.www.pacificdisaster.net/pdnadmin/data/original/unhabitat_2008_gendering_land_tools.pdfUN-HABITAT. 2009. Gender evaluation criteria for large-scale land tools. Nairobi.

UN-HABITAT. 2008a. Secure Land Rights for All.Nairobi.

UN-HABITAT. 2011. Designing and evaluating land tools with a gender perspective: training package for land professionals. Nairobi.

UNICEF. no date. Communication for development.www.unicef.org/cbsc/index_42346.html

UN Women. 2011. *Progress of the world's women 2011–2012: In pursuit of justice.* New York. http://progress.unwomen.org/pdfs/en-report-progress.pdf WLLA. 2010a. *The impact of national land policy and land reform in Uganda.* Accra, Women's Land Link Africa.

Wehrmann, B. 2003. Women's access to and control over land in African and South Asian countries. *Landnutzung und Landentwicklung* [Land Use and Development], 44: 83–89.

Wehrmann, B. 2011. *Land use planning. Concept, tools and applications.* Eschborn, Germany, GIZ. www2.gtz.de/dokumente/bib-2011/giz2011-0041en-land-use-planning.pdf

Whitehead, A. &Tsikata, D. 2003. Policy discourse on women's land rights in sub-Saharan Africa: The implications of the re-turn to the customary. *Journal of Agrarian Change, 3*(1–2): 67–112.

WLLA. 2010a. *The impact of national land policy and land reform in Uganda.* Accra, Women's Land Link Africa.

WLLA. 2010b. *The impact of national land policy and land reform on women in Zambia.* Accra.

WOLREC. 2011. *Women's access to land and household bargaining power: a comparative action research project in patrilineal and matrilineal societies in Malawi. Research Report No. 9.* Rome, ILC, Women's Legal Resource Centre. www.landcoalition.org/sites/default/files/publication/959/wlr_9_malawi.pdf

World Bank. 2005. *Gender issues and best practices in land administration projects: A synthesis report.* Washington, DC. http://siteresources.worldbank.org/intard/resources/gender_land_fulltxt.pdf

World Bank/FAO/IFAD. 2009. Gender in agriculture sourcebook. Washington, DC and Rome.

Yahya, S. 2002. In G. Payne, ed. *Land rights and innovation: improving tenure security for the urban poor.* London, ITDG Publishing.

World Bank/FAO/IFAD. 2009a. Module 4: Gender issues in land policy and administration. *In Gender in agriculture sourcebook*. Washington, DC and Rome.

World Bank/FAO/IFAD. 2009b. Module 10: Gender and natural resources management. *In Gender in agriculture sourcebook*. Washington, DC and Rome.

YWAP. 2011. *Assisting Kayole widows in gaining control to family land, Kenya*. Research Report No. 10. Rome, ILC, Young Widows Advancement Program. www.landcoalition.org/sites/default/files/publication/951/wlr_10_ywap.pdf

Further resources for gender-equitable governance of land tenure

The following links to additional resources on the Internet that may be relevant sources of further guidance and information in the pursuit of responsible gender-equitable governance of land tenure are not comprehensive as the pool of available resources is constantly expanding. The list focuses on Web sites that emphasize gender and land specifically rather than on more general land-related Web sites.

FAO Gender and Land Rights Database
This provides data and information on gender and land rights issues from across the globe. www.fao.org/gender/land rights

Global Land Tool Network (GLTN) Gendering Land Tools Mechanism
This provides information about land tools and tool development, including work on the Gendering Land Tools Mechanism and on gender evaluation criteria for land tools.
www.gltn.net/index.php/land-tools/cross-cutting-tools/gender

International Land Coalition (ILC) Land Portal Gender Topic
This has a regularly updated Gender Topic page as well as links to many relevant papers, discussions and resources.
http://landportal.info/topic/gender

International Land Coalition (ILC) Women's Access to Land
The new section on Women's Land Rights includes links to many sources of information and resources on women and land, news items and research produced through ILC programmes. www.landcoalition.org/global-initiatives/womens-land-rights/womens-land-rights

Land for African Women
The Website of the African Women's Land Rights Conference held in Nairobi in 2011 includes conference papers, other resources, a blog and regularly updated news items. http://landforafricanwomen.org/

Landesa Center for Women's Land Rights
This provides access to information about Landesa's projects on women and land, a blog and links to research and resources.
http://www.landesa.org/women-and-land/

Land Rights in Africa
This contains many resources on land and gender and is regularly updated.
www.mokoro.co.uk/land-rights-in-africa

USAID Land Tenure and Property Rights Portal
This is regularly updated and includes access to downloadable products, including issues briefs, tools and training materials, some of which address gender.
http://usaidlandtenure.net/

International instruments in support of gender-equitable governance of land tenure

The Voluntary Guidelines to Support the Progressive Realization of the Right to Adequate Food in the Context of National Food Security (2005)

Guidelines 2.5, 3.5, 3.8, 3.9, 7.4, 8.3, 8.6 and 8.10 specifically relate to the promotion of inclusive, fair and non-discriminatory policies of special relevance to gender and governance of tenure; Guideline 8.6 calls for States to promote women's full and equal participation in the economy including through implementation of gender-sensitive legislation providing women with the right to inherit and possess land and other property; Guideline 8.10 calls for States to take measures to promote and protect the security of land tenure, especially with respect to women and poor and disadvantaged segments of society, through legislation that protects the full and equal right to own land and other property, including the right to inherit.

The Pinheiro Principles (2005)

Principles 11 to 14 on land restitution policy specify that these must be non-discriminatory and recognize the claims of men and women and boys and girls; Principle 18 entitles subsidiary claimants including spouses to make restitution claims on the same basis as the primary claimant.

The Protocol to the African Charter on Human and People's Rights, on the Rights of Women in Africa (2003)

Article 6 provides for women to have the right to acquire and manage their own property during marriage; Article 7 provides for the rights of women to have an equitable share of joint property in marriage in the case of divorce, separation or annulment; Article 21 provides for the rights of widows to an equitable share in inheritance on the death of their husbands, including the right to remain in the matrimonial home, and for the rights of men and women to inherit equitable shares of their parents' property; Article 16 provides for women to have the right to equal access to housing; Article 19 promotes women's access to and control over productive resources and guarantees their right to property as part of their right to sustainable development.

The Habitat Agenda Goals and Global Plan of Action (1996)

Provides for adequate shelter for all, including equal access to land and housing finance without discrimination, and for policies and practices to promote equal participation of women in human settlement planning and decision-making.

The World Food Summit Action Plan (1996)

Objective 1.3 requires governments to ensure gender equality and women's empowerment through introducing and enforcing gender-sensitive legislation providing women with secure and equal access to and control over land and other productive resources.

The Beijing Declaration (1995)

Article 35 restates the determination of governments to ensure women's equal access to economic resources, including land and credit.

Agenda 21 (1992)

Paragraph 3.8 calls for governments to strengthen legal frameworks for land management, access to land and landownership, particularly for women; Paragraph 32.6 requires governments to protect and formalize women's tenure and use of land and their rights to land.

The Convention on the Rights of the Child (1989)

Article 2 protects children against the effects of discrimination against their parents and thereby effectively protects their mothers from gender-based discrimination.

The International Covenant on Civil and Political Rights (1966)

Articles 2, 3 and 27 provide for rights to be enjoyed without discrimination according to gender; Article 23.4 specifies that the property rights of women in marriage and at its dissolution should be protected as well as those of men.

The International Covenant on Economic, Social and Cultural Rights (1966)

Article 3 provides for men and women to have equal rights and equal enjoyment of their rights.

The Universal Declaration of Human Rights (1948)

Article 2 upholds the entitlement of every human being to enjoy human rights and fundamental freedoms without discrimination; Article 7 entitles every human being to equal protection under law; Article 16 entitles men and women to equal rights during marriage and at its dissolution.